Advances in Material Property Characterization using Small Punch Tests

I0751076

The small punch test (SPT) is useful to calculate changes in the tensile and fracture properties of structural materials during the service life of the materials of plant components. This book compiles advances in the development of correlations to calculate mechanical properties of the materials using SPT data. New correlations have been developed using hybrid methodology involving analytical and experimental data. The newly developed correlations have been tested conducting case studies on SPT and pre-cracked/notched SPT (p-SPT) specimens. The eventual applications of all the new correlations have been demonstrated by conducting a real-life case study involving degradation of structural material from ductile to semi-brittle state due to aging.

Features:

- Presents exclusive material on the remnant life assessment of in-service materials using SPTs.
- Assesses the fracture toughness of ductile materials using the experimentally measured biaxial fracture strain.
- Provides new equations to calculate the yield and ultimate stresses of copper and titanium alloys using measured SPT data.
- Explores functions to correlate the load-displacement data of p-SPT specimens with fracture properties.
- Includes case studies with direct relevance to the degradation of plant materials.

This book is aimed at researchers, professionals, and graduate students in materials science and engineering, mechanical property characterization and testing, and small-scale experimentation. It is expected that the advanced methodology presented in this book to evaluate changes in the properties of aged materials during the service life using SPT data are useful to designers for safety evaluation and also to calculate the remaining service life of industrial components for life extension studies.

Advances in Material Property Characterization using Small Punch Tests

Edited by
Bijan Kumar Dutta

CRC Press is an imprint of the
Taylor & Francis Group, an informa business

First edition published 2025
by CRC Press
2385 NW Executive Center Drive, Suite 320, Boca Raton, FL 33431

and by CRC Press
4 Park Square, Milton Park, Abingdon, Oxon, OX14 4RN

CRC Press is an imprint of Taylor & Francis Group, LLC

ISBN: 9781032315751 (hbk)
ISBN: 9781032315768 (pbk)
ISBN: 9781003310372 (ebk)

DOI: 10.1201/9781003310372

Typeset in Times
by Newgen Publishing UK

Contents

Preface

It is a worldwide practice to extend the life of existing plants due to outstanding safety records during operation. It is generally found that the accumulated material damage in various plant components over their service life is generally limited due to advances made in design procedures, material development, fabrication techniques, skillful plant operations, carefully scheduled maintenance procedures, and the implementation of stringent safety measures. The low accumulation of material damage at the end of the design life inspires plant operators to evaluate the possibility of extending the life of plant components. This also helps to overcome the socio-economic issues associated with the installation of a new plant.

One of the important tasks to carry out in the life extension exercise for existing plant components is the determination of the mechanical properties of aged materials at the end of their service life. Such inputs are particularly important for, for example, chemical plants, nuclear plants, and high temperature power plants, in which material damage is significant due to adverse service conditions. Material characterizations associated with such studies need to be obtained using an extremely limited quantity of degraded materials. Due to such limitations, it is not possible to fabricate conventional tensile and fracture toughness test specimens of the required dimensions following ASTM E8/E8M and ASTM E1820 standards. To overcome such difficulties, the small punch test (SPT) was developed in the early 1980s. This test is primarily used to assess changes in the tensile and fracture properties of structural materials during the service life of the materials of plant components, and to calculate updated safety margins and the remaining life. The test is performed on a small specimen of the order of 3 mm diameter with 0.25 mm thickness, or a 8 mm square with 0.5 mm thickness. The scrapping of such a small quantity of material from an existing plant component in service is generally considered to be non-destructive testing without affecting the life of the component. The SPT specimens are tested in a dedicated high-precision machine especially meant to test such miniaturized specimens. The output of such tests is the load vs. load-line displacement data of the specimen until fracture.

It is an involved task to convert such load-displacement data to the mechanical properties of the materials using empirical/semi-empirical correlations. The development of improved correlations is an active area of research all over the world. Continued research work in this area is aimed at improving the predictive capabilities of the correlations to obtain realistic material properties using SPT load-displacement data. An extensive research program has been conducted by a group of doctoral students over several years to develop such correlations, which have the capability to predict material properties with improved accuracies. Such research work has not only led to the publication of many peer-reviewed journal papers but also delivered a number of doctoral level theses. A number of new methodologies adopted to develop improved correlations can also be used by other researchers in the future for similar purposes. The present book is an effort to compile the outcome of such advanced research for the benefit of graduate students, doctoral students, and practicing engineers engaged in life management studies of plants.

Dr. B.K. Dutta
Editor

About the Editor

B.K. Dutta, Bachelor of Engineering (National Institute of Technology, Raipur), Master of Technology (Indian Institute of Technology, Kanpur), Doctor of Philosophy (Indian Institute of Technology, Bombay), served at the Department of Atomic Energy during the period Aug 1977–Dec 2018. Subsequently, he was associated with departmental programs as a Raja Ramanna Fellow and Academic Consultant during the period Jan 2019–Jul 2023. As a Distinguished Scientist, Head Human Resource Development Division (Bhabha Atomic Research Centre – BARC), and Dean (Homi Bhabha National Institute – HBNI), he served the department as a researcher, academician, and engineering practitioner. The primary thrust of Dr. Dutta's research is linked to the structural safety of Indian nuclear reactors. As an academician, he successfully supervised 12 Ph.D. students and 15 M.Tech. students, in addition to providing day-to-day guidance to his departmental colleagues. He has been associated with the HBNI from its inception and served the institute in various capacities, including as Dean and Vice Chancellor (Officiating). As an engineering practitioner, he participated in addressing a number of technical issues related to nuclear and other conventional facilities, which helped the management in decision-making.

Dr. Dutta has published 300+ research papers, including 120+ peer-reviewed international journal articles and a book chapter. He also served as guest editor of number of special issues of various journals. He was former president and presently lifetime Advisory Board member of the International Association for Structural Mechanics in Reactor Technology (IASMiRT – USA) and a member of different professional societies. He is a recipient of the Technical Excellence Award DAE, Homi Bhabha Award for Science and Technology DAE, Distinguished Faculty Award HBNI and Lifetime Achievement Alumnus Award NIT Raipur. He served as chair/member of the selection boards of prestigious Indian organizations for many decades. He was elected as a Fellow of the Indian National Academy of Engineering in 2004.

Contributors

Bijan Kumar Dutta
Former Distinguished Scientist, DAE and
Dean, Homi Bhabha National Institute
Training School Complex
Anushakti Nagar, Mumbai 400094, India

Jayanta Chattopadhyay
Reactor Safety Division
Bhabha Atomic Research Centre
Trombay, Mumbai 400085, India

Shyam Rao Ghodke
Accelerator and Pulse Power Division
Bhabha Atomic Research Centre,
Trombay, Mumbai 400085, India

Pradeep Kumar
Structural Mechanics Department
Design at Plasser India Pvt. Ltd
Karjan, Gujarat 392210, India

Taslim D. Shikalgar
Former Doctoral Student, Homi Bhabha National Institute
Training School Complex
Anushakti Nagar, Mumbai 400094, India

1 Introduction

B.K. Dutta

1.1 OVERVIEW

One of the important tasks of a design engineer is to ascertain the safety of industrial plant components during the entire service period, even under extreme loading conditions. Some examples of extreme loading conditions are emergency conditions related to plant operations, earthquakes, hurricanes, and other natural or man-made disasters. Structural integrity assessment is further vital for plants in which the mechanical properties of materials change significantly during their service life, such as chemical plants, nuclear plants, offshore structures, and biomedical plants. One needs to know the current material properties to carry out safety analyses at any point in time during the service period. For this purpose, a small quantity of material may be scrapped directly from an in-service plant component. However, one needs to ensure that the structural integrity of the component, and hence serviceability, is not affected by such a scrapping process. Therefore, the quantity of scrapped material available to evaluate a change in mechanical properties should be very limited. Hence, it is not feasible to conduct the testing of such in-service materials using ASTM standard specimens. To overcome this difficulty, it is required to involve advanced test procedures, such as miniaturized specimen testing.

A similar situation is to conduct the mechanical testing of irradiated materials of a nuclear power plant using ASTM standard specimens. The structural materials of a nuclear reactor are exposed to high doses of irradiation by nuclear particles during their service life. Because of this, the mechanical properties of such materials change significantly and the materials lose ductility. To quantify such changes in material properties, material coupons are placed in the nuclear reactor core at different locations. These coupons are taken out from time to time and tested to determine the change in the mechanical properties of the materials caused by irradiation. There are difficulties in irradiating ASTM standard specimens using this approach, such as the availability of limited space in a nuclear reactor core to accommodate multiple ASTM standard specimens; the non-uniform gradients of the irradiation dose in a thick specimen, and hence, the damage; and radiation

DOI: 10.1201/9781003310372-1

exposure to personnel involved in post-irradiation testing. Hence, there is again a need to use miniaturized specimens in such conditions. The change in material properties caused by irradiation is particularly required to carry out safety studies of nuclear plant components and also a mandatory input to carry out life extension studies for existing plants.

The above examples demonstrate the need to develop dependable miniaturized specimen testing methodology. One such testing methodology is the small punch test (SPT).

1.2 FRAMEWORK OF THE SPT

The SPT was primarily proposed to characterize engineering materials of industrial plant components where an adequate quantity of material is not available to fabricate conventional tensile and fracture test specimens as per ASTM E8/E8M and ASTM E1820 standards. The SPT is performed on a miniaturized circular or square specimen with a thickness of 0.25–0.50 mm. Such a small quantity of material can be made available by scrapping an in-service plant component without losing its useability. The European Code of Practice (EUCoP) on Small Punch Testing to determine tensile and fracture properties was published by CEN in 2006 and revised in 2007 [1]. The tests involve clamping a specimen between two dies and then punching the specimen with a rigid spherical ball until fracture, as shown in Fig. 1.1. The measured load vs. load-line displacement (LLD) data are recorded during the experiment. Such data may then be used to calculate parameters such as the initial stiffness, yield load, maximum load, maximum displacement at failure, and area under the load-displacement curve. These experimental parameters can then be used to calculate several mechanical properties of the material, such as the yield stress, ultimate stress, hardening coefficients, biaxial fracture strain, and fracture toughness. The SPT has many benefits, including the following: (a) it can be used to monitor the change in material properties of in-service plant components, (b) the experiment is simple to perform, (c) it can be used to study the change in the properties of aged materials exposed to hazardous environments that are difficult

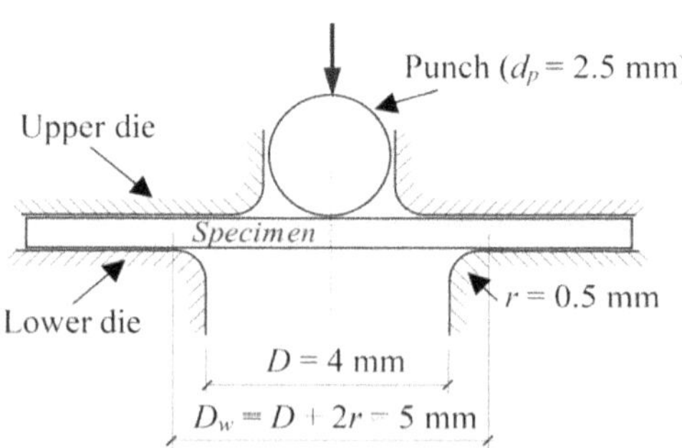

FIG. 1.1 Schematic diagram of an experimental setup for an SPT

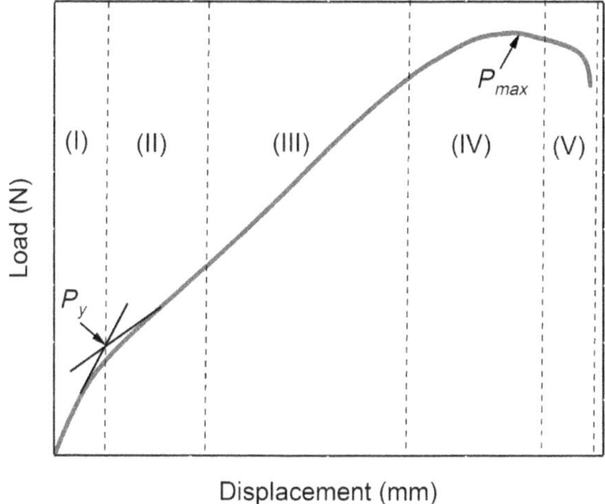

FIG. 1.2 Different stages of deformation of the SPT specimen during tests

to handle, such as nuclear irradiation, harmful chemicals, and toxin organics, (d) it can be used to conduct the low-cost screening of newly procured materials using random sampling, and (e) it provides a quick assessment of the properties of new alloys under development.

Different stages of deformation of an SPT specimen during loading are shown in Fig. 1.2. These stages are (I) elastic bending, (II) plastic bending, (III) plastic bending and membrane stretching, (IV) geometrical softening due to material damage, and (V) final failure. In the figure, (a) P_y is the initial localized plastic strain load and can be correlated to the yield stress, (b) P_{max} is the maximum load and can be correlated to the ultimate stress of the material, (c) the final displacement at fracture can be correlated to the biaxial fracture strain, (d) the segment of the curve between P_y and P_{max} can be correlated to hardening coefficients, and (e) the area under the entire load-displacement curve can be correlated to the specimen energy at fracture. These five parameters are used to calculate the corresponding material mechanical properties by employing semi-empirical/empirical correlations. Limitations of the available empirical correlations are presented in the next section.

Over the last few decades, efforts have been made by many researchers to quantify the fracture properties of engineering materials using conventional SPT specimens employing different empirical correlations. Such equations correlate material fracture toughness with the biaxial strain at fracture. These correlations are primarily based on the best-fit curve on large sets of experimental data for a particular material [2]. The assessment of material fracture properties using these correlations without having an initial crack in the SPT specimen lacks some degree of confidence. To circumvent this difficulty, several authors [3–5] introduced an initial crack/notch in the SPT specimens, as shown in Fig. 1.3. These specimens are primarily used to estimate the fracture toughness of engineering materials.

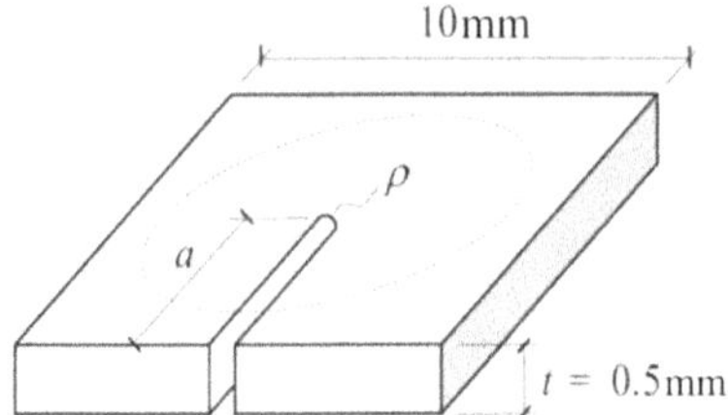

FIG. 1.3 Schematic diagram of a pre-cracked SPT specimen

These SPT specimens are called pre-cracked/notched SPT (p-SPT) specimens. The fracture parameters, that is, J-integral and crack tip opening displacement (CTOD), can be obtained using the experimental data of p-SPT specimens.

1.3 AVAILABLE CORRELATIONS AND LIMITATIONS

The accuracy of material properties determined by SPT data depends on the correlations used for this purpose. Voluminous literature is available on the development of such correlations. However, the available correlations have certain limitations that readers should be aware of. Some of these are summarized below.

(a) Two widely used correlations in SPT literature are to calculate the yield and ultimate stresses using SPT data. These correlations in the literature are mostly developed using the experimental data of ferrous materials and thus only applicable to ferrous materials.
(b) A well-established methodology is needed to generate the entire stress-strain data of an engineering material using SPT data.
(c) Another useful correlation in the literature is to calculate the biaxial strain using SPT data. The available correlation uses two empirical coefficients whose values are determined using experimental data and proposed to be constants. The dependence of these empirical coefficients on the geometry and material parameters of the SPT specimen needs to be investigated.
(d) The correlation employed to calculate fracture toughness using the biaxial fracture strain was established using the experimental data of ferrous materials. The availability of a correlation that is material independent will improve the usability of the SPT methodology.
(e) A set of correlations denoted by η-functions is extensively used to correlate the two global fracture parameters (LLD and crack mouth opening displacement (CMOD)) with the two crack tip local parameters (J-integral and CTOD) of a fracture specimen. Similarly, another correlation denoted by the m-factor is used to correlate the two crack tip local parameters J-integral and CTOD. The forms of the η-function and m-factor are not available in the literature for p-SPT specimens.

(f) One of the most important applications of SPT specimens is to assess the change in the mechanical properties of engineering materials during the service period. In such cases, the available quantity of aged materials is very limited and not sufficient to prepare ASTM standard specimens. Hence, SPT specimens are very useful under such conditions. A detailed case study demonstrating the step-by-step application of SPT specimens to quantify the properties of a real-life engineering aged material is useful for practitioner engineers.

1.4 CHAPTERWISE CONTENT

The objective of Chapter 1 is to introduce the SPT and its importance, the limitations of the available literature, and the chapterwise content of the present book. Chapter 2 summarizes the literature on the SPT, the test methodology, a description of load-displacement data, useful correlations available to calculate the mechanical properties of engineering materials using SPT data, modeling methodologies for material damage, and the usefulness of the two parameters 'η' and 'm' in fracture mechanics.

The objective of Chapter 3 is to demonstrate a comprehensive analytical methodology to develop new correlations to calculate the yield and ultimate stresses of any material using SPT specimen data. In the literature, such correlations are developed mostly for ferrous alloys using a statistical analysis of the experimental data. As the availability of experimental data for developing such correlations of a material of the user's interest is not always guaranteed, this chapter presents an alternative method to develop such important correlations analytically.

Chapter 4 presents an improved correlation between the minimum specimen thickness and the corresponding punch displacement of an SPT specimen during testing. Such a correlation is available in the literature using two empirical coefficients, whose values are taken to be constants. A detailed study is presented to determine the dependency of these two coefficients on the geometrical and material parameters of the SPT specimen. This study reveals the dependency of this correlation on specimen thickness. The new empirical correlations are presented based on this study.

The objective of Chapter 5 is to present a new procedure to calculate fracture toughness using SPT specimen data. Such a procedure is available in the literature correlating the biaxial fracture strain with the crack initiation toughness. Multiple empirical correlations are available in the literature for this purpose. However, these have been developed using the experimental data of a particular material and hence are applicable to those specific materials. The present chapter shows a new methodology to generate a material-independent generalized correlation for this purpose.

Chapter 6 deals with η-functions and m-factors. These two specimen-dependent parameters are used extensively in fracture mechanics. There are two global and two crack tip local parameters in a fracture specimen. The global parameters are

LLD and CMOD. The local parameters are J-integral and CTOD. There are four η-functions that correlate four combinations of global and local parameters. The η-functions are specimen dependent and the expressions are available in the literature for commonly used fracture specimens. In addition, the m-factors that correlate the two local parameters (J-integral and CTOD) of a fracture specimen are also useful to the fracture mechanics community. It may be noted here that the expressions of the η-functions and m-factors of p-SPT specimens are not available in the literature. The purpose of the present chapter is to demonstrate numerical procedures to derive expressions of the η-functions and m-factors of p-SPT specimens. Readers may also note that the procedure presented in this chapter can also be used to derive such expressions for other fracture specimens.

The primary objective of Chapter 7 is to acquaint readers with a methodology for calculating the material J-R curve in the case of the availability of a very limited quantity of material and where the material is unsafe to handle physically. There are two methodologies to calculate the J-R curve of an engineering material: (a) the experimental method using ASTM standard fracture specimens and (b) finite element analysis (FEA) incorporating material damage models, such as the GTN model and cohesive zone model. In this chapter, a new hybrid technique is presented for this purpose that is primarily applicable to SPT data. In the case of the availability of a limited quantity of hazardous material, the new method is very useful. This method does not need a large quantity of material and also does not involve advanced FEA with intricate material properties as inputs. The development and application of the hybrid technique are presented in this chapter. The chapter also deals with the experimental and numerical validation of this technique.

The mechanical and fracture properties of engineering materials change during their service life due to metallurgical and micro-structural changes. Chapter 8 deals with real-life case studies to describe the application of SPT specimens to quantifying mechanical property degradation during the service life of an aged material. To demonstrate the procedure, oxygen-free electronic (OFE) copper is taken as the candidate material, which has wide applications in the nuclear industry, particularly in fusion nuclear reactors and linear accelerators (LINAC). The SPT specimens of the OFE copper are subjected to electron irradiation to simulate aging effects. The degraded SPT specimens are then tested, and the load-displacement data thus generated are used to quantify changes in the mechanical and fracture properties. The Holloman power law and cohesive zone parameters are then calculated by carrying out finite element parametric studies of SPT data. Subsequently, the J-R curve of the aged OFE copper is calculated employing these parameters to a three-point bending specimen. This real-life case study demonstrates the importance of assessing changes in the properties of aged materials with a service life to carry out a safety evaluation and determine the remaining service life of industrial components.

1.5 CHAPTER CLOSURE

The objective of the present chapter is to acquaint readers with the SPT methodology, the importance of different correlations used to calculate the mechanical properties of engineering materials using SPT data, the limitations of such correlations and scope for improvement, and the chapterwise content of the present book. The next chapter will help readers to become familiar with the current state of the literature on SPTs.

REFERENCES

[1] Small Punch Test Method for Metallic Materials, CWA 15627. Part A: A Code of Practice for Small Punch Creep Testing and Part B: A Code of Practice for Small Punch Testing for Tensile and Fracture Behaviour, Documents of CEN WS21, Brussels, 2007

[2] Mao X and Takahashi H, Development of a further miniaturized specimen of 3 mm diameter for TEM disk (φ 3 mm) small punch tests, Journal of Nuclear Materials, 150, 42–52, 1987

[3] Martínez-Pañeda E, García T E, Rodríguez C, Fracture toughness characterization through notched small punch test specimens, Materials Science and Engineering A, 657, 422–430, 2016

[4] Alegre J M, Cuesta I I, Barbachano H L, Determination of the fracture properties of metallic materials using pre-cracked small punch tests, Fatigue and Fracture of Engineering Materials and Structures, 38,104–112, 2015

[5] Cuesta I I, Rodriquez C, Belzunce F J, Alegre J M, Analysis of different techniques for obtaining pre-cracked/notched small punch test specimens, Engineering Failure Analysis, 18, 2282–2287, 2011

2 Current Literature on Small Punch Test Methodology

B.K. Dutta, Pradeep Kumar, Taslim D. Shikalgar, and S.R. Ghodke

2.1 OVERVIEW

Extensive studies on small punch tests (SPTs) have been carried out over many decades. Such studies comprise types of specimens, fixtures, and test methodologies, and semi-empirical/empirical correlations to analyze SPT data to determine the material properties, such as the yield stress, ultimate stress, biaxial fracture strains, and fracture toughness. The first SPT was performed by Manahan et al. [1] and Huang et al. [2]. The objective was to estimate the mechanical properties of a material that is available in a very limited quantity. The SPT is performed on a small circular or square specimen by applying a mechanical load to one of the surfaces of the specimen by means of a hard spherical ball. The experimental load-line displacement (LLD) of the specimen as a function of loads is recorded. These data are then used to calculate the material properties. The present chapter is devoted to summarizing the current state of the literature related to the SPT.

2.2 SPT SPECIMENS, TEST FIXTURES, AND ANALYSIS OF TEST RESULTS

2.2.1 Types of SPT Specimens and Test Fixtures

The SPT fixture assembly is composed of a lower and upper die, a hard spherical ball, and a mechanical drive to move the hard spherical ball. Voluminous literature is available on such tests [1–72]. In addition to uncracked SPT specimens, some researchers have also proposed a pre-cracked/notched SPT (p-SPT) specimen to determine the fracture toughness of a material. The most commonly used specimen geometry and fixtures were initially proposed by Manahan et al. [1] and Baik et al. [3, 4] to predict the mechanical and fracture properties of a material. Table 2.1 shows different types of proposed SPT specimens and the fixture dimensions. The fixture assembly has two important dimensions: (a) the bore diameter of the lower die through which the specimen deforms (d_2), and (b) the bore diameter of the

DOI: 10.1201/9781003310372-2

TABLE 2.1
SPT specimen and die dimensions recommended by different authors

Reference	SPT specimen dimension (mm)	SPT die dimension (mm)
Manahan et al. [1]	$Ø = 3; t_0 = 0.25$	$d_2 = 1.5; d_1 = 1$
Baik et al. [3, 4]	$10 \times 10 \times 0.5$	$d_2 = 3.4; d_1 = 2.4$
Mao et al. [5]	$Ø = 3; t_0 = 0.25$	$d_2 = 1.5; d_1 = 1$
Mao et al. [5, 6]	$10 \times 10 \times 0.5$	$d_2 = 3.4; d_1 = 2.4$
Misawa et al. [8]	$10 \times 10 \times 0.25$	$d_2 = 4; d_1 = 2.4$
Rasche et al. [24]	$Ø = 8; t_0 = 0.5$	$d_2 = 4; d_1 = 2.5$
Garcia et al. [43]	$10 \times 10 \times 0.5$	$d_2 = 4; d_1 = 2.5$

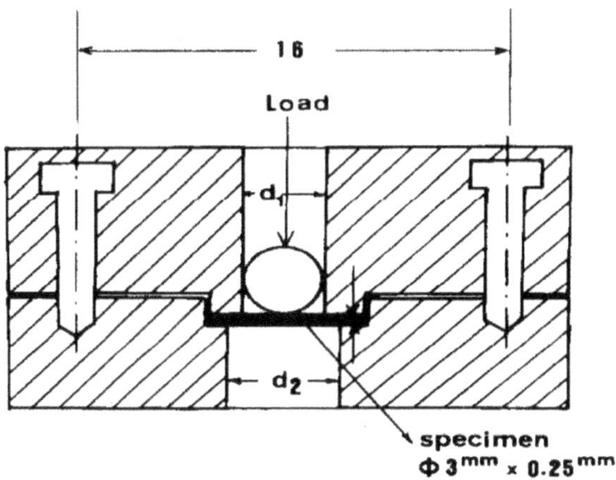

FIG. 2.1 Schematic diagram of an SPT specimen and fixture assembly [5]

upper die and the diameter of the spherical ball (d_1). Other important dimensions are the fillet radius (r*) of the lower die, clamping area, and specimen dimensions. According to the CEN code of practice on the "small punch test method for metallic materials", a fillet radius (r*) of 0.2 mm is recommended for a 3 mm diameter disk specimen [41]. Another important recommendation is that the hardness of the SPT fixtures should not be less than 55 HRC, as recommended in [41]. The salient dimensions of the recommended fixtures are shown in Fig. 2.1.

The shape of the thin SPT specimen may be circular or square. Table 2.1 shows the recommended shapes and sizes of SPT specimens from different researchers. Table 2.2 shows the recommended dimensions of p-SPT specimens, the salient dimensions of the die, and the shape of different notch profiles. The two types of recommended notch profiles are (a) a through-thickness notch and (b) a part-through thickness notch with a predefined depth to thickness ratio (a/t). Fig. 2.2

TABLE 2.2
Different types of p-SPT specimen and die dimensions recommended by different authors

Reference	P-SPT specimen dimension (mm)	P-SPT die dimension (mm)	Through thickness notch length, 'a' (mm)	Notch depth to thickness ratio (a/t)
Cuesta et al. [28]	20 × 20 × 1	$d_2 = 8$; $d_1 = 5$	---	0.2–0.6
Cardenas et al. [29]	10 × 10 × 0.5	$d_2 = 4$; $d_1 = 2.5$	---	0.4
Yifei Xu et al. [30]	Ø = 10; $t_0 = 1$	$d_2 = 4$; $d_1 = 2.5$	---	0.5
Cuesta et al. [31]	10 × 10 × 0.5	$d_2 = 4$; $d_1 = 2.5$	5 and 5.4	---
Garcia et al. [32]	10 × 10 × 0.5	$d_2 = 4$; $d_1 = 2.5$	---	0.3
Martinez-Paneda et al. [33]	10 × 10 × 0.5	$d_2 = 4$; $d_1 = 2.5$	---	0.3
Alegre et al. [45]	10 × 10 × 0.5	$d_2 = 4$; $d_1 = 2.5$	4-6	---

(a–c) shows procedures to fabricate a p-SPT specimen with different notch profiles. Readers may note that, due to the requirement of a few milligrams of material needed to fabricate an SPT specimen, this test may be treated as a non-destructive test. It is possible to scrap such an amount of small material from an in-service component to ascertain aging effects on the material without affecting the useability of the plant component.

2.2.2 SPT Experiment and Analysis of Load-Displacement Data

The SPT is generally performed at a crosshead speed of 0.2 mm/min. The accuracy of the measured experimental data strongly depends on the axial alignments of different experimental components. For this purpose, the following three precautions should be taken during fabrication: (a) the precise alignment of the axis of the die with the axis of the punch, (b) close geometrical tolerances during the fabrication of the punch, lower die, and upper die, and (c) uniform thickness of the specimen with very close tolerances [5–7].

The SPT should be performed with a high precision measurement system to measure punch loads and punch displacements very accurately. One of the following three procedures to measure the displacement may be adopted for this purpose: (a) reading the punch displacement from the inbuilt data measurement system of the machine, (b) measurement of the punch displacement using a dedicated clip gauge, or (c) measurement of the central deflection on the opposite face of the SPT specimen employing a linear variable differential transducer (LVDT).

Interpretations of the mechanical deformation in different regions of an experimental load-displacement curve are shown in Fig. 2.3 [24]. A typical SPT load-displacement plot for a ductile material may be divided into five regions. To understand the deformation behavior of each region, Onat [46] performed a

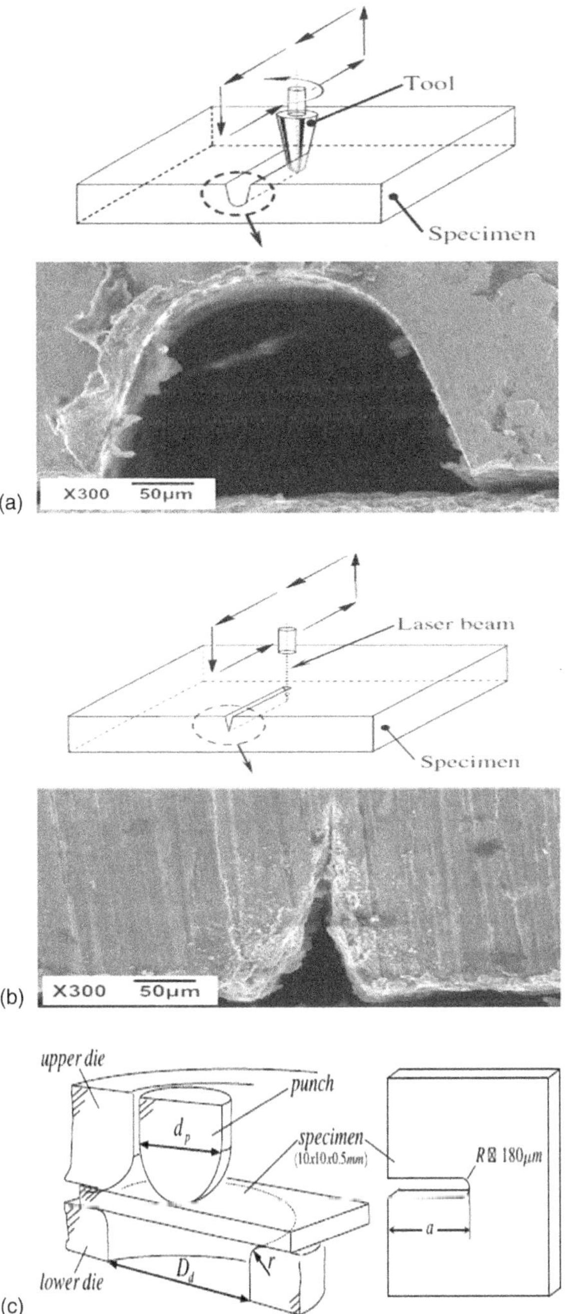

FIG. 2.2 Fabrication procedure of different types of notch profile in a p-SPT specimen: (a) mechanical micro-machining, (b) laser-induced micro-machining notch profile to fabricate a part-through thickness notch with different depth to thickness ratios, and (c) schematic diagram of the machining of a through thickness notch profile [31, 44]

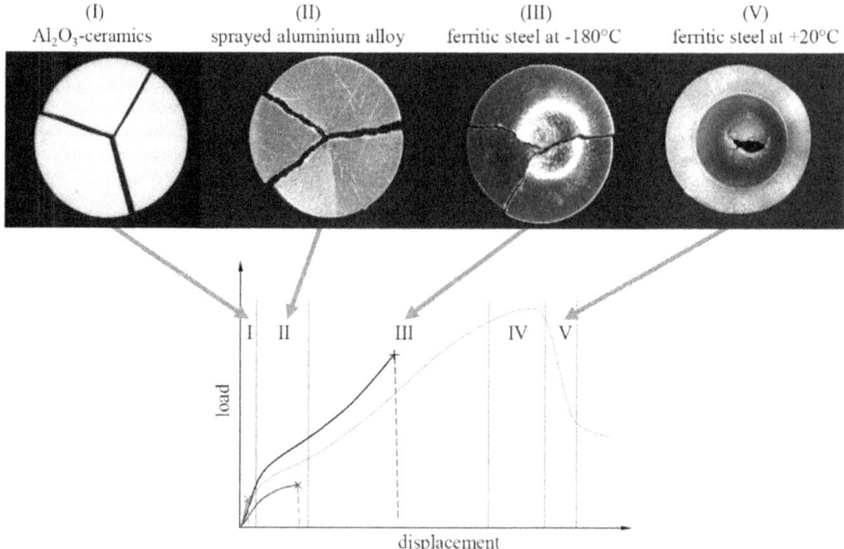

FIG. 2.3 Schematic SPT load-displacement diagram showing different stages of the deformation pattern until fracture. The shapes of the plots and the displacement at failure imply material toughness. This ranges from purely brittle (e.g., alumina ceramics) to fully ductile (e.g., ferritic steel) [24]

numerical simulation of the SPT using a perfect rigid plastic model. During the initial loading, the small punch specimen elastically deforms, as represented by region-I. Subsequently, the deformation of the SPT specimen is in good agreement with the plastic bending of the specimen. This is represented by region-II. In region-III, the deformation of the SPT specimen lies within plastic bending and the membrane stretching (or bulging) regime, during which the slope increases due to work hardening of the material. The phenomena of geometrical softening and the damage mechanism develop all through region-IV. On approaching the maximum load, the slope of the plot decreases due to the accumulation of significant material damage caused by the formation of voids or microcracks, depending on the type of material being tested. Subsequently, the specimen loses its load carrying capacity represented by region-V. Readers may note that the presence of different regions in the SPT load-displacement plot depends on the type of material being tested, that is, brittle, semi-brittle, semi-ductile, or ductile, as illustrated in Fig. 2.3.

An important correlation between the thickness at the thinnest location of the SPT specimen during testing and its central displacement is shown in eq. (2.1); this was derived by Mao et al. [5] based on the large number of SPT data on structural steels. Fig. 2.4 shows the plots of such data [5]:

$$t/t_0 = \exp[-\beta(\delta/t_0)^p], \tag{2.1}$$

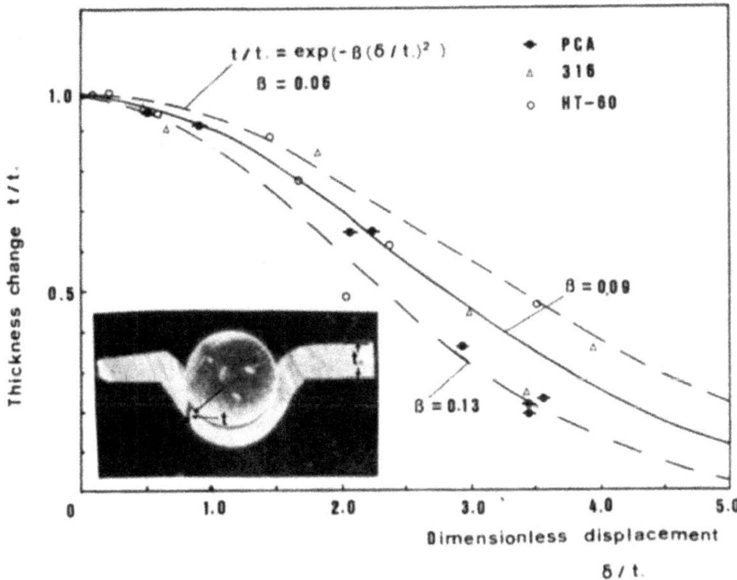

FIG. 2.4 Correlation between the minimum thickness and load line displacement of an SPT specimen during loading [5]

where t is the current minimum thickness of the SPT specimen at any location and δ is the current LLD during loading. The symbol t_0 denotes the initial thickness of the specimen. Mao et al. [5] proposed the values of β and p as 0.09 and 2, respectively, for the SPT disk specimen with diameter φ = 3 mm. This was calculated using a large number of experimental data. It may be emphasized here that the values of such constants were derived using the experimental data of different structural steels and their alloys. Hence, the applicability of these two empirical constants to other materials needs to be examined. Kulkarni et al. [47] opined that the deformed shape of an SPT disk specimen can be represented by a group of tensile specimens placed in the radial direction of the disk at different circumferential angles, as shown in Fig. 2.5. It may also be noted that the experimental load vs. displacement plot of a p-SPT specimen has a shape and deformation regimes very similar to those shown in Fig. 2.6.

2.2.3 Effect of Different Parameters on the Load vs. LLD of the SPT Specimen

a) Effect of the Shape of the Punch

The plastic and fracture behavior of a material during SPT have been found to be sensitive to the shape of the punch [4]. It has been shown that a flat punch leads to a flow curve similar to those found in uniaxial tensile tests. This is because of the absence of plastic bulging. A cone-shaped punch does not show a clear

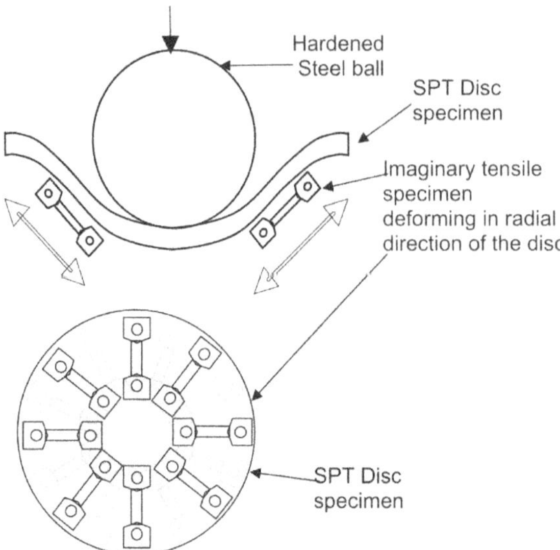

FIG. 2.5 Schematic diagram showing how the deformation of an SPT specimen represents the combined deformation of a group of tensile specimens subjected to biaxial strain placed circumferentially [47]

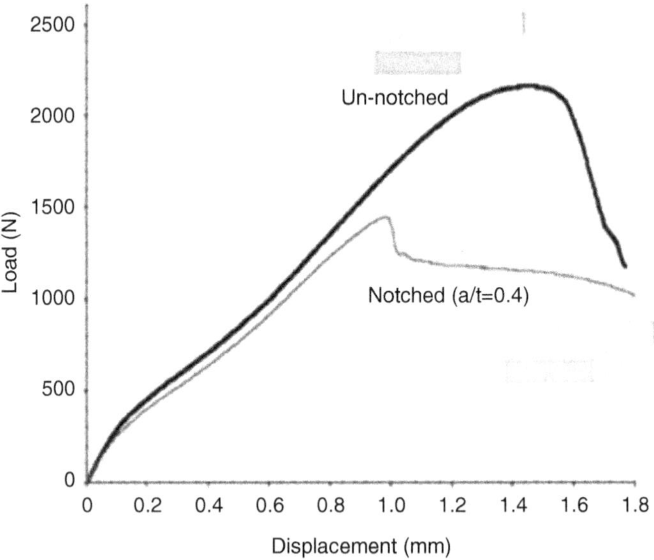

FIG. 2.6 Schematic diagram showing comparison of shape of load vs. punch displacement plots between an uncracked SPT specimen and a pre-cracked SPT specimen

fracture point in the load-displacement plot because plastic bulging during the test is small. A flat punch with rounded corners shows a clear fracture point in the load vs. displacement plot, however without any plastic bulging in the SPT specimen. A spherical punch test shows a clear fracture point in the load vs. displacement plot, along with broader plastic bulging. This happens due to an increase in the contact area during the deformation between the punch and the specimen. The bulge ductility at the point of fracture of the specimen is considered to be a fracture parameter because it can be correlated with the fracture toughness of the materials [5–23, 48].

b) Effect of the Test Temperature

A decrease in the test temperature leads to a reduction in the ductility of ferritic steels. Such a phenomenon is visible during SPTs. The fracture point in the load vs. LLD plot shifts from regime-V to regime-I based on the test temperature. Voluminous information is available on this phenomenon in different references [4, 8, 17, 21, 24, 49, 50].

c) Effects of In-Service Aging

The aging effects on a material subjected to different plant environments, such as nuclear irradiation, hydrogen embrittlement, stress corrosion, and grain boundary precipitation, can be observed during the SPT of a material. Such SPT results are investigated in great detail in different references [6, 9, 11, 15, 16, 21]. In general, such aging effects lead to an increase in the yield and maximum loads during the tests. The decrease in ductility is measured by calculating the biaxial fracture strain. This is also manifested by the reduction in the area under the load-displacement plots.

d) Effect of Geometrical and Material Parameters

The load vs. LLD of an SPT specimen is a strong function of the geometrical and material parameters. These have been investigated by many authors in the literature [18, 24, 42, 51–54]. The maximum load increases as the specimen thickness, diameter of the ball indenter, and coefficient of friction between the ball indenter surface and the specimen surface increase. Similarly, the maximum load also increases as the flow stress of the specimen material increases. However, the maximum load decreases as the grain size of the specimen material increases, and it has no effect on the change in the modulus of elasticity.

Peñuelas et al. [55] and Cuesta et al. [56] studied the effect of elastic-plastic parameters and Gurson–Tvergaard–Needleman (GTN) damage parameters in different regions of the load-displacement plot of steels. The experimental load-displacement curve of the SPT specimen was simulated using finite element analysis (FEA) to evaluate the macro-mechanical parameters (E, v, σ_y, n, and K) and GTN micro-mechanical parameters (ε_n, f_n, f_c, and f_F) of the steels. The effect of these parameters in different regions of the load-displacement plot are shown in Fig. 2.7 [55, 56].

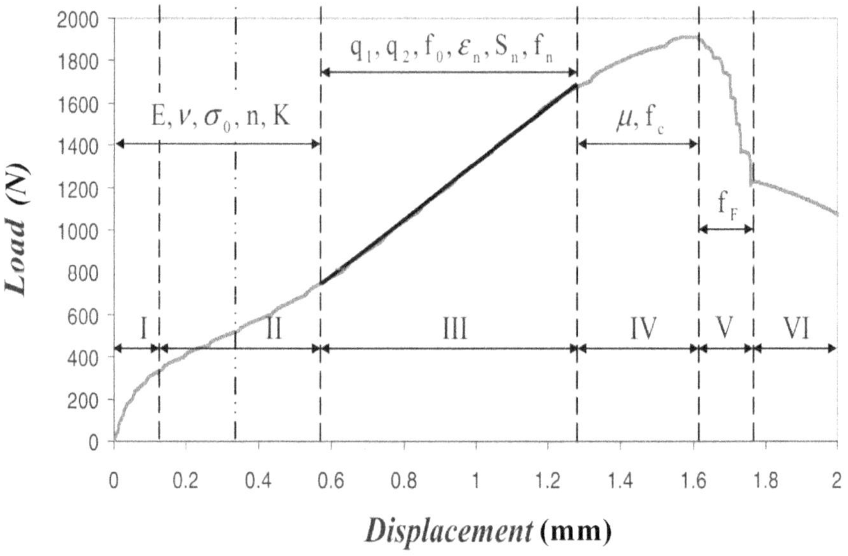

FIG. 2.7 Effect of the macro- and micro-mechanical parameters that govern each region of the SPT load vs. punch displacement curve of structural steels [55]

e) Effect of Cold Working

Cuesta and Siegl [57, 58] also studied the effect of the tensile and compressive cold working process on the SPT load-displacement data. They observed that the value of yield load P_y increases as the compressive plastic strains increase due to cold working and hence, the calculated yield strength of the material also increases. On the other hand, the maximum load P_{max} and hence the calculated ultimate strength of the material is not affected with the amount of tensile and compressive cold working. Another important parameter is the LLD at the maximum load P_{max}. Cuesta and Siegl observed that, with an increase in cold working, the value of P_{max} decreases and hence the area under the load vs. LLD data. This implies a loss of ductility of the specimen material with an increase in cold working.

f) Effect of the Punch Velocity

Manahan et al. [1] studied the influence of a punch velocity of the order of 10^{-4} mm/s ($\approx 10^{-5}$ strain/sec) to 10^{-2} mm/s ($\approx 10^{-3}$ strain/sec) on small punch experimental data for a variety of materials. Mao et al. [6, 7] also conducted a small punch experiment at a punch velocity of the order of 0.2 mm/min ($\approx 10^{-4}$ strain/sec). In general, it has been observed that the maximum load P_{max} increases as the punch velocity increases.

2.3 CORRELATIONS TO DETERMINE THE MECHANICAL PROPERTIES OF MATERIALS USING SPT DATA

2.3.1 Determination of the Yield Load and Yield Stress

Different researchers proposed methodologies to calculate the yield load applying SPT specimen experimental data [5, 41, 43, 59–61]. One of the widely used schemes was suggested by Mao et al. [5]. In this scheme, the yield load is identified as the intersection point between two tangents defined in the elastic bending regime (region-I) and plastic bending regime (region-II) of the load-displacement plot. This is denoted in this chapter by P_{y_Mao}. The CEN code of practice [41] proposed a slightly different criterion. In this case, P_y is identified by a vertical projection of the same intersection point of the two tangents on the experimental curve, which is denoted in this chapter by P_{y_CEN}. Other researchers [59, 60] identified the P_y load as the intersection point between the SPT curve and a straight line parallel to the initial slope of the graph with an offset displacement of $t/10$ or $t/100$, where t is the initial thickness of the specimen. These are denoted by $P_{y\_t/10}$ or $P_{y\_t/100}$, respectively. The most recent proposal to identify the value of P_y was given by Lacalle et al. [61]. It is calculated by the load corresponding to the first inflexion point located in region-I and is denoted by P_{y_inf}. Fig. 2.8 shows all these proposals graphically for the benefit of readers.

A number of empirical correlations have been proposed by different authors to calculate the yield stress (σ_{ys}) of materials [5, 43, 52, 53, 57, 59, 60, 62–67, 69] using the value of yield load P_y. Most researchers have agreed on a linear correlation between σ_{ys} and P_y/t^2 as follows:

$$\sigma_{ys} = \alpha_1 \frac{P_Y}{t^2} + \alpha_2 \tag{2.2}$$

where t is the initial thickness of the SPT specimen and α_1 and α_2 are empirical constants. Table 2.3 shows the values of α_1 and α_2 proposed by different researchers. An appropriate definition of P_y needs to be used from Fig. 2.8 when using the values of α_1 and α_2 shown in Table 2.3.

2.3.2 Determination of Ultimate Stress using SPT Data

The ultimate tensile stress (UTS) of the SPT specimen material is also calculated by empirical correlation using experimental data. Different forms of the correlation have been suggested by different researchers [5, 43, 50, 52, 53, 59, 60–70]. These are shown as follows:

$$\sigma_{ut} = \beta_1 \frac{P_m}{t^2} + \beta_2 \tag{2.3}$$

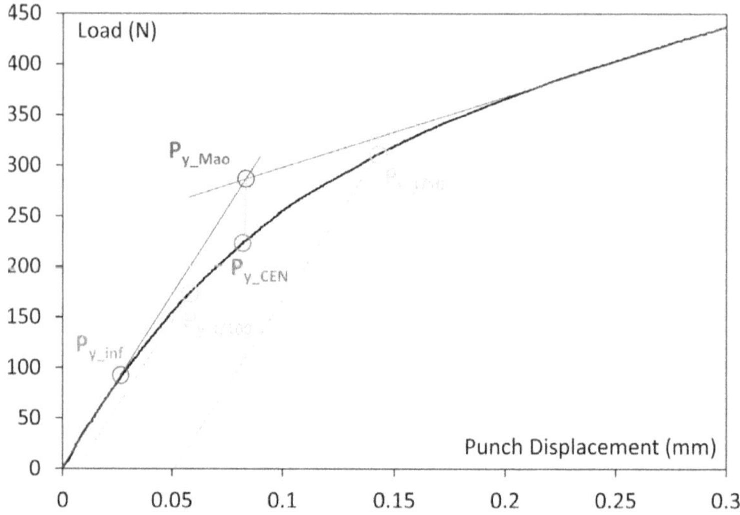

FIG. 2.8 Schematic representation of various proposals to determine yield load P_y using the SPT load-displacement plot [43]

TABLE 2.3
Values of empirical constants in the proposed correlations to calculate the yield stress using the yield load of SPT data

	Correlation coefficients	
Reference	α_1	α_2
Mao et al. [5]	0.360	-
Garcia et al. [43]	0.442	-
I.I. Cuesta et al. [57]	0.700 (tensile pre-strain)	-107.90
I.I. Cuesta et al. [57]	0.418 (compressive pre-strain)	-6.20
Rodríguez et al. [59]	0.380	-
Xu et al. [62]	0.477	-
Cheon et al. [63]	0.630	-
Ruan et al. [64]	0.413	149
Isselin et al. [65]	0.540	-

$$\sigma_{ut} = \beta_1' \frac{P_m}{t} + \beta_2' \tag{2.4}$$

$$\sigma_{ut} = \beta_1'' \frac{P_m}{(t^* dm)} + \beta_2'', \tag{2.5}$$

TABLE 2.4
Values of empirical constants in the proposed correlations to calculate the ultimate stress using the maximum load of SPT data

Reference	Empirical correlation coefficients					
	β_1	β_2	β'_1	β'_2	β''_1	β''_2
Mao et al. [5]	0.130	-320	-	-	-	-
Garcia et al. [43]	0.065	268.81	0.129	286.7	0.277	-
Kumar et al. [50]	0.13	6	-	-	-	-
Bruchhausen et al. [53]	0.093	-11.86			0.326	-27.04
Rodríguez et al. [59]	0.077	218	-	-	-	-
Ruan et al. [64]	0.077	218	-	-	-	-
Altstadt et al. [69]	0.09	-			0.25	-

where $\beta_1, \beta_2, \beta'_1, \beta'_2, \beta''_1, \beta''_2$ are empirical constants, P_m is the maximum load, t is the initial thickness of the specimen, and d_m is the LLD at the maximum load in SPT data. The values of the empirical constants as suggested by different researchers are shown in Table 2.4.

2.4 DETERMINATION OF THE FRACTURE PROPERTIES OF A MATERIAL USING SPT DATA

2.4.1 Past Investigations

Baik et al. suggested the use of SPT data to generate material fracture data [3, 4]. This work created all-round interest in the scientific community. Mao et al. [5–7] conducted SPTs on different structural steels using either 3 mm diameter disk specimens with 0.25 mm thickness or 10 mm × 10 mm × 0.5 mm square specimens. The objectives were to develop empirical correlations to calculate the fracture properties. The estimation of fracture toughness was performed in [8] for cryogenic steels at low temperatures using SPT data. Additionally, fracture toughness degradation was investigated on neutron irradiated 2.25Cr-1Mo ferritic steel in [6, 9]. More research on the prediction of fracture properties using the SPT data has been carried out by various researchers [3–35] for different materials and test conditions to standardize the tests. Geary et al. performed experiments on three structural steels and one high strength offshore steel [10]. Saucedo-Muñoz et al. developed a fracture toughness correlation for austenitic stainless steels [11]. Guan et al. calculated the fracture toughness of CrMo low alloy steels after long-term in-service operation [12]. Rodríguez et al. performed fracture characterization on a hot rolled API X-70 steel plate and its heat affected zone [13]. Bulloch compiled the studies carried out by various researchers to evaluate fracture toughness (J_{1C}) correlations using the biaxial fracture strain and specimen energy [14]. Several authors have performed the SPT and Charpy impact test to evaluate the ductile-to-brittle transition temperature (DBTT) [9, 15–17]. The objectives were to develop empirical correlations between the DBTT measured by the Charpy impact test and SPT data.

Hybrid methods involving experiments and numerical analyses to determine the mechanical and fracture properties using SPT data have been attempted by several researchers [18–27]. In general, a two-dimensional (2D) axisymmetric finite element model of the SPT specimen is employed to determine load vs. displacement data. Partheepan et al. [27] performed the FEA simulation of a dumb bell miniature specimen and used a neural network model to evaluate fracture toughness. In recent years, several researchers have also proposed using p-SPT specimens [28–35].

There are several ASTM special technical publications (STPs) that deal with the SPT that provide a comprehensive overview of the development of the SPT for the evaluation of material mechanical and fracture properties. Some of these publications are, "The Use of Small-Scale Specimens for Testing Irradiated Materials STP 888" [36], "SPT Techniques Applied to Nuclear Reactor Vessel Thermal Annealing and Plant Life Extension STP 1204" [37], and "Small Specimen Test Techniques: Fourth Volume STP 1418, 5th Volume STP 1502 and 6th volume 1576" [38–40].

2.4.2 Determination of the Biaxial Fracture Strain using SPT Data

The biaxial fracture strain $\bar{\varepsilon}_{qf}$ is an important SPT output parameter. A number of researchers have tried to correlate the biaxial fracture strain with the fracture toughness values of the SPT specimen material. Chakrabarty et al. [71] proposed a correlation to calculate the biaxial strain in a circular plate specimen using an analytical method. The equivalent biaxial strain equation was based on the membrane stretch theory and is given by

$$\bar{\varepsilon}_q = \ln\left(\frac{t_0}{t}\right) \tag{2.6}$$

One needs to measure the minimum thickness at the fracture in an SPT specimen to calculate the biaxial fracture strain using eq. (2.6). The measurement of such data experimentally is an involved task and prone to errors. To circumvent the problem associated with the measurement of the minimum thickness, Mao et al. [5] proposed the following equation to correlate the LLD of the specimen with the corresponding minimum thickness of the SPT specimen during deformation:

$$t/t_0 = exp\left[-\beta\left(\delta/t_0\right)^p\right]. \tag{2.7}$$

Using eqs. (2.6) and (2.7), one can obtain the biaxial equivalent strain at the fracture of the SPT specimen as

$$\bar{\varepsilon}_{qf} = \ln(t_0/t_f) = \beta(\delta^*/t_0)^p \tag{2.8}$$

where t, t_0, t_f, δ^*, δ, and $\bar{\varepsilon}_{qf}$ are the minimum thickness in the SPT specimen during loading, initial thickness, minimum thickness at the fracture, LLD at the fracture, current LLD, and biaxial equivalent fracture strain, respectively. Using a large number of experimental data on structural steels, Mao et al. [5] suggested the values of β and p as 0.09 and 2, respectively. These values are applicable for the SPT disk specimen of 3 mm diameter and 0.25 mm thickness.

2.4.3 Determination of the Fracture Energy of an SPT Specimen

The SPT fracture energy E_{SP} is defined as the area under the load vs. punch displacement curve up to the point of fracture during the test. This is calculated using

$$E_{SP} = \int_0^{\delta^*} F(\delta)d\delta \tag{2.9}$$

The function $F(\delta)$ can be approximated using multivariate polynomial regression up to the point of failure. The SPT specimen fracture energy is often correlated with the plane-strain fracture toughness J_{Ic} and also with the equivalent fracture strain $\bar{\varepsilon}_{qf}$ [14].

2.4.4 Determination of the Fracture Toughness of SPT Specimen Material

A significant amount of research has been carried out to develop empirical correlations to calculate fracture toughness using SPT data [5–18, 43]. The proposed correlations primarily correlate the biaxial fracture strain ($\bar{\varepsilon}_{qf}$) calculated using SPT data with the fracture toughness (J_{Ic}) of the material as follows:

$$J_{Ic} = C_1\bar{\varepsilon}_{qf} \pm C_2 \tag{2.10}$$

where C_1 and C_2 are empirical constants. These correlations are primarily developed based on a large set of experimental data. The equivalent biaxial fracture strain ($\bar{\varepsilon}_{qf}$) may be determined for engineering materials using SPT data by testing a specimen as small as a transmission electron microscopy disk with a 3 mm diameter and 0.25 mm thickness. In miniaturized specimen techniques, the size of the specimen is too small to meet the size requirement of valid material fracture toughness J_{Ic}. Therefore, it is important to determine a correlation between the equivalent biaxial fracture strain ($\bar{\varepsilon}_{qf}$) and the material fracture toughness (J_{Ic}) for design purposes. Mao et al. conducted a large set of experiments on different ferrous materials to develop a correlation between the biaxial equivalent fracture strain ($\bar{\varepsilon}_{qf}$) obtained

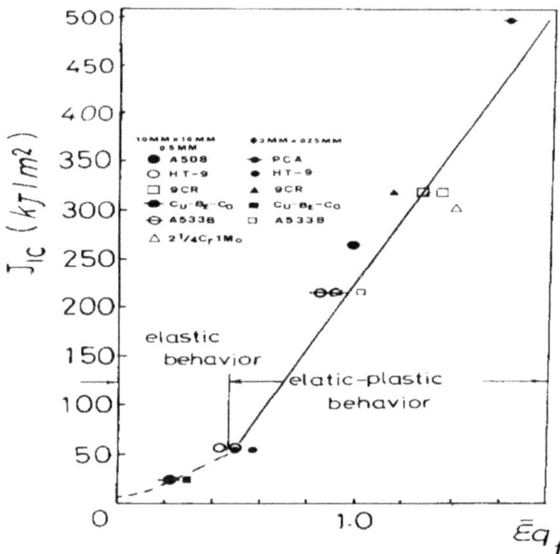

FIG. 2.9 Correlation between the measured J_{IC} and $\bar{\varepsilon}_{qf}$ for a ductile material [5]

using SPT data and the fracture toughness (J_{Ic}) obtained by testing a standard fracture specimen [5]. The developed correlation shows an approximately linear correlation between these two parameters, as shown in Fig. 2.9, when the equivalent fracture strain is larger than 0.20. Therefore, from Fig. 2.9, it can be concluded that the experimental evidence is generally consistent with the theory proposed by Bayoumi [48], which showed the linear dependency of J_{Ic} on the biaxial fracture strain ($\bar{\varepsilon}_{qf}$) for ductile materials.

Fig. 2.10 shows the plot of different correlations proposed by various authors for this purpose. It may be noticed that correlations vary significantly because of different definitions of fracture initiation toughness used by various authors. Table 2.5 shows the values of empirical constants C_1 and C_2 in these correlations.

2.5 MODELING OF MATERIAL DAMAGE LEADING TO FRACTURE

2.5.1 GTN Model

One of the widely used formulations to model the damage of ductile materials during deformation is popularly known as the GTN model [74, 75, 79]. Voluminous literature is available on the GTN model [74–81]. A brief description is provided here as a ready reference for readers. This model is also used in subsequent chapters to develop advanced correlations and case studies on SPT specimens.

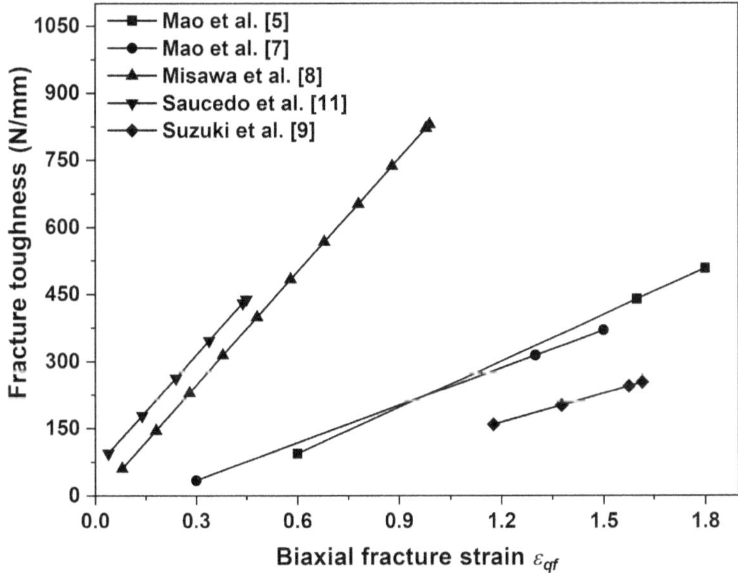

FIG. 2.10 Empirical correlations between the biaxial fracture strain and fracture toughness proposed by different researchers

TABLE 2.5
Proposed values of the empirical constants of different correlations to calculate fracture toughness using the equivalent biaxial fracture strain

	Proposed values of empirical correlation constants	
Reference	C_1	C_2
Mao et al. [5, 6]	345	-113
Misawa et al. [8]	845	-7
Suzuki et al. [9]	215	-94
Geary et al. [10]	154	286
Saucedo et al. [11]	840	62
Guan et al. [12]	276.77	0.5207
Rodríguez et al. [13]	433	-159
Garcia et al. [43]	1695	-1320

In the GTN model, the yield function is defined as

$$\varphi(\sigma_m, \sigma_0, f) = \left(\frac{\sigma_{eq}}{\sigma_0}\right)^2 + 2f^* q_1 \cosh\left(\frac{3q_2\sigma_m}{2\sigma_0}\right) - \left(1 + q_3\left(f^*\right)^2\right) = 0, \quad (2.11)$$

where the macroscopic von Mises stress is $\sigma_{eq} = \left(\frac{3}{2}s_{ij}s_{ij}\right)^{1/2}$, where (s_{ij}) are the stress deviators and σ_0 is the matrix level flow stress of the material, f is the void volume fraction, and q_1, q_2, q_3 are fitting parameters suggested by Tvergaard [76, 77]. The function f^* was introduced in [75] to model the rapid loss of the stress carrying capacity of the material caused by void coalescence as observed during the experiments. This function is expressed as

$$f^* = \begin{cases} f & for\, f \leq f_C \\ f_c + \dfrac{f_U^* - f_C}{f_F - f_C}\left(f - f_C\right) & for\, f > f_C. \end{cases} \tag{2.12}$$

The function becomes effective once the void volume fraction f exceeds a critical value f_C. The complete loss of load-carrying capacity occurs at $f = f_F$ $i.e\, f^*(f_F) = f_U^* = 1/q_1$.

The evolution law for the void volume fraction is given by

$$\dot{f} = \dot{f}_{growth} + \dot{f}_{nucleation}. \tag{2.13}$$

From the assumption of isochoric plasticity and conservation of mass, the rate of change of the void volume fraction is proportional to the rate of volumetric plastic strain and is given by

$$\dot{f}_{growth} = (1 - f)\delta_{ij}\,\dot{\varepsilon}_{ij}^p, \tag{2.14}$$

where δ_{ij} is the Kronecker delta function and $\bar{\varepsilon}_{df}^p$ is the plastic strain tensor. The nucleation rate is controlled either by the plastic strain or the stress and is assumed to follow a normal distribution as follows:

$$\dot{f}_{nucleation} = A\dot{\bar{\varepsilon}}^p, \tag{2.15}$$

where $\dot{\bar{\varepsilon}}^p$ is the equivalent plastic strain rate and

$$A = \frac{f_N}{S_n\sqrt{2\pi}}\exp\left(-\frac{1}{2}\left(\frac{\bar{\varepsilon}^p - \varepsilon_n}{S_n}\right)^2\right), \tag{2.16}$$

where ε_n is the mean strain and S_n is the standard deviation. These values are suggested as $\mathcal{E}_n = 0.3$, $S_n = 0.1$ [78]. The suggested values of q_1, q_2, q_3 are 1.5, 1.0, 2.25, respectively. The material input parameters to apply the GTN model are f_0, f_n, f_c, and f_f, which are the initial void volume fraction, void volume fraction at nucleation, void volume fraction at coalescence, and void volume fraction at fracture, respectively.

To compute J-initiation as well as the entire J-R curve [73] close to the experimental values using the GTN model, a phenomenological form of the q_2 parameter was suggested in the GTN constitutive model [81] to analyze structures under the presence of a crack. The modified form of the q_2 parameter is

$$q_2 = 1 + q_{2a} e^{-(r/l_c)/q_{2b}}. \tag{2.17}$$

It has an exponential spatial variation near the crack tip and two empirical constants. These additional constants help analysis to compute the J-initiation, as well as the entire J-R curve close to the experimental data. The suggested values of the two additional constants are $q_{2a} = 0.3$ and $q_{2b} = 8$, and r is the radial distance from the crack tip and l_c is a characteristic length generally taken as 0.1 mm. Readers will be shown different case studies involving the GTN model in subsequent chapters. The determination of the GTN material parameters f_0, f_n, f_c, and f_f using small punch experimental data and an artificial neural network are also explained in detail.

2.5.2 Cohesive Zone Model (CZM)

Similar to the GTN model, the CZM is also used to address the loss of load carrying capacity of an engineering material due to damage. In the CZM, a traction–separation law (TSL) is employed to describe the loss of load-bearing capacity of a material as a function of the separation between two layers. This is independent of the physical damage phenomenon that occurs in the material. As a result, it may be used to model both ductile, semi-ductile, semi-brittle, and brittle material damage. In contrast to the GTN model, the CZM is reasonably independent of the mesh size, which makes it particularly appealing. The phenomenological CZM is the most user-friendly model in comparison to other material damage models available in the literature. The CZM is preferred over other models because of its numerical resilience, mesh insensitivity, and the requirement of only two model parameters.

The CZM can be used to model material damage leading to fracture in a variety of metals and non-metals over a wide range of lengths and time scales, such as metals and alloys, ceramics, polymers, concrete, wood, fiber-reinforced materials, glass, and rock [90]. To formulate the CZM, Schwalbe [82] and Dugdale [83] proposed a strip-yield model based on the assumption that a cohesive force prevents a crack from propagating, as shown in Fig. 2.11. In this earlier model, the magnitude of the cohesive force is assumed to be equal to the material yield strength σ_{ys}. Strain hardening is not considered, implying that the material should behave in an elastic-ideally plastic manner. The formation of a physically unrealistic singularity at the crack tip is avoided as the local stress is restricted up to the yield stress of the material. Subsequently, this model is used to determine the crack tip opening displacement (CTOD), as well as the opening profile of the crack.

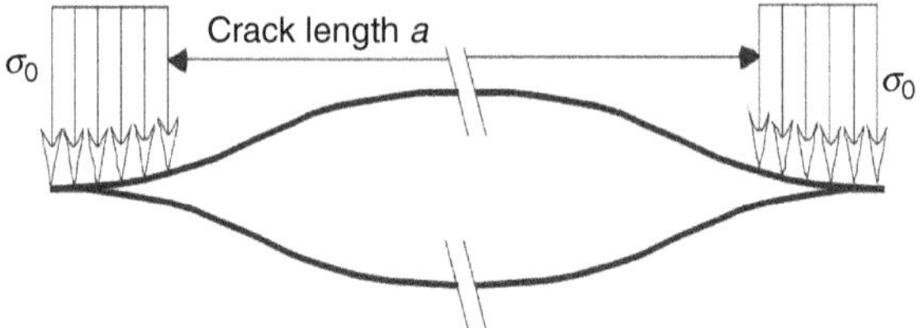

FIG. 2.11 Schematic representation of the Dugdale model

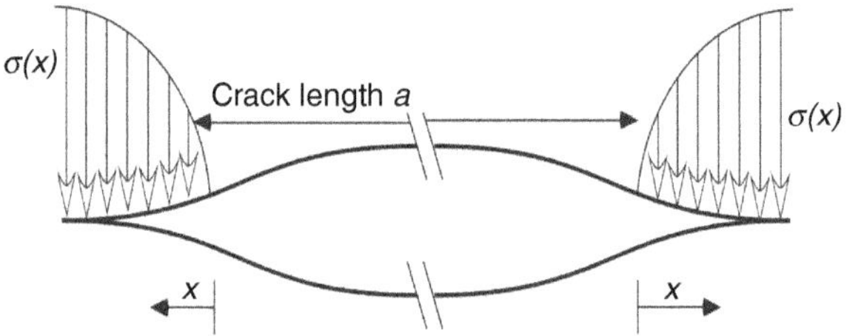

FIG. 2.12 Schematic representation of the Barenblatt model

Later, the CZMs made use of the concept of Barenblatt's work [84], as shown in Fig. 2.12. In such models, the yield strength is replaced by a cohesive law to simulate the de-cohesion of the atomic lattices. The plastic zone is thus replaced by a process zone, which subsequently leads to damage and fracture. It is worth noting that traction in Barenblatt's model is stated to be a function of the distance from the crack tip, whereas traction in the advanced cohesive models is defined as a function of separation within the cohesive zone. Material degradation and separation occur in a distinct plane, which is represented by cohesive elements embedded in the continuum parts that represent the test piece. Hillerborg et al. [85] employed the cohesive model to characterize the damage behavior of concrete. This was the first time that cohesive models were applied to characterize the fracture behavior of concrete.

Needleman, Tvergaard, and Hutchinson pioneered studies for a very significant class of engineering materials. Needleman [86] performed the first microdamage analysis in ductile materials, while Tvergaard and Hutchinson investigated the first macroscopic crack extension in ductile materials [87]. In subsequent studies [88, 91], comparisons of the results of the cohesive model of ductile materials with the experimental results were carried out.

Traction–Separation Law (TSL): A TSL correlates traction T with separation δ representing the separation distance between the two surfaces of a cohesive element. The TSL is used to describe the constitutive behavior of the cohesive model. A cohesive element fails once the separation distance reaches a material-specific critical separation value δ_0. In such a situation, the resultant stress in the

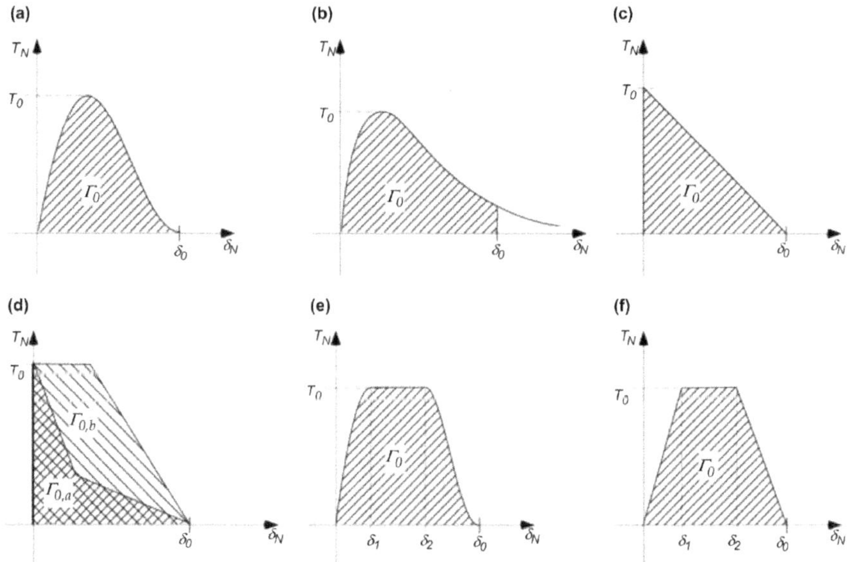

FIG. 2.13 Typical TSLs proposed by different authors: (a) Needleman [86], (b) Needleman [89], (c) Hillerborg [85], (d) Bazant [90], (e) Scheider [91], and (f) Tvergaard and Hutchinson [87]

element is zero. Another material property required to specify the TSL is the cohesive strength T_0. This is the maximum stress that can be attained during the separation process. Fig. 2.13 shows a list of commonly used shapes of the TSL.

In the literature, a TSL with a finite initial stiffness and exponential form, as shown in Fig. 2.13 (a), is frequently employed for ductile materials. Such a variation also represents the loss of material strength with a softening curve approaching zero traction asymptotically, as shown in Fig. 2.13 (b). Other material specific shapes of the TSL used by different researchers are shown in Fig. 2.13 (c–f).

The two material input parameters, δ_0 and T_0, are adequate for modeling the complete separation process for a given shape of TSL. In practice, cohesive energy Γ_0 is also used in place of a critical separation displacement δ_0. The following equation shows an expression used to calculate cohesive energy as a function of T and δ; this is the energy required to generate a unit area of the fracture surface:

$$\Gamma_0 = \int_0^{\delta_0} T(\delta)\, d\delta. \tag{2.18}$$

The cohesive model can also be used to model different failure mechanisms. This also means that, in principle, the model can be used for any material and for any mode of fracture independent of the loading conditions. A single layer of cohesive elements is adequate in FEA for modeling the entire fracture process in the case in which the fracture path is known in advance. In subsequent chapters, readers will be conversant with the methodology used to calculate cohesive parameters using SPT data. A strategy to model the complete fracture of an SPT specimen observed during an experiment by employing cohesive elements will also be described. The cohesive parameters calculated using the SPT data of an aged material may further be used to calculate the material fracture toughness by analyzing ASTM standard fracture specimens. This subject will also be dealt with subsequently.

2.6 SIGNIFICANCE OF η-FUNCTIONS IN FRACTURE MECHANICS

The principle of fracture mechanics deals with the four important parameters of a cracked specimen. Out of these, two parameters are termed 'global parameters': (a) the LLD and (b) the crack mouth opening displacement (CMOD). Both of these parameters can be measured directly during an experiment. The other two parameters, termed 'local crack tip parameters', are not measurable experimentally: (a) the CTOD and (b) J-integral. Local parameters are employed to characterize a crack, such as crack initiation and crack growth. All the four parameters are shown in Fig. 2.14 schematically for the benefit of readers.

Local crack tip parameters can be related to global parameters, which are measured during fracture experiments. For such purposes, η-functions are used, which correlate local parameters with global parameters. To develop η-functions, the crack tip local parameters J and CTOD are expressed as the sum of elastic and plastic components [92]:

$$J = J_{el} + J_{pl} \tag{2.19}$$

$$\delta = \delta_{el} + \delta_{pl}, \tag{2.20}$$

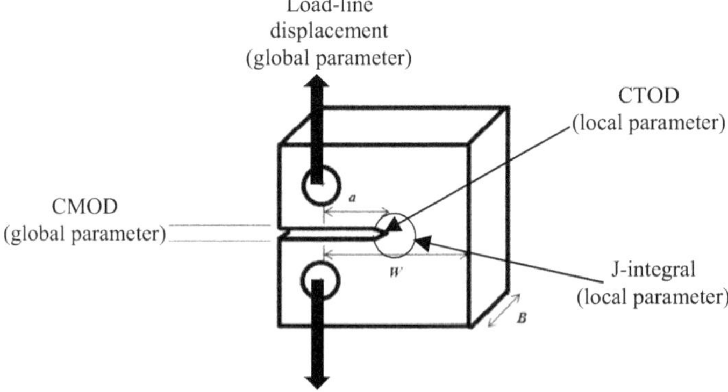

FIG. 2.14 Schematic representation of global and local parameters in a cracked specimen

where J_{el} and J_{pl} are the elastic and plastic components of the J-integral, respectively. Similarly, δ_{el} and δ_{pl} are the elastic and plastic components of the CTOD, respectively. The elastic components of the above equations are expressed as

$$J_{el} = \frac{(1 - v2) K_I^2}{E} \tag{2.21}$$

$$\delta_{el} = \frac{(1 - v2) K_I^2}{m E \sigma_{ys}}, \tag{2.22}$$

where σ_{ys}, E, and v are the yield stress, Young's modulus, and Poisson's ratio, respectively. The parameters K_I and m are the stress intensity factor and an empirical plastic constraint factor, respectively. Details of the m-factor are presented in the subsequent section of the present chapter. The following four correlations [93] can be used to calculate the plastic components J_{pl} and δ_{pl}:

$$J_{pl} = \frac{\eta_{j,pl}^{LLD} A_{pl}^{LLD}}{b B_N} \tag{2.23}$$

$$J_{pl} = \frac{\eta_{j,pl}^{CMOD} A_{pl}^{CMOD}}{b B_N} \tag{2.24}$$

$$\delta_{pl} = \frac{\eta_{\delta,pl}^{LLD} A_{pl}^{LLD}}{b B_N \sigma_F} \tag{2.25}$$

$$\delta_{pl} = \frac{\eta_{\delta,pl}^{CMOD} A_{pl}^{CMOD}}{b B_N \sigma_F}, \tag{2.26}$$

where b, B_N, and σ_F are the remaining ligament $(W - a)$, net specimen thickness, and flow stress of the material, respectively. A_{pl}^{LLD} and A_{pl}^{CMOD} are the plastic components of the strain energy during the deformation of the specimen at any instant during loading. A_{pl}^{LLD} and A_{pl}^{CMOD} are determined by calculating the area under the curve between load vs. LLD_{pl} or between load vs. $CMOD_{pl}$, respectively. The four η-functions correlate the two global parameters with the two local parameters as follows:

(i) The $\eta_{J,pl}^{LLD}$ function correlates J_{pl} with LLD_{pl}.
(ii) The $\eta_{J,pl}^{CMOD}$ function correlates J_{pl} with $CMOD_{pl}$.
(iii) The $\eta_{\delta,pl}^{LLD}$ function correlates $CTOD_{pl}$ with LLD_{pl}.
(iv) The $\eta_{\delta,pl}^{CMOD}$ function correlates $CTOD_{pl}$ (δ) with$CMOD_{pl}$.

Readers may note the following two points:

a. The η-functions depend on the type of fracture specimens, such as compact tension (CT) and three-point bending (TPB). The correlations to determine the η-functions for different types of fracture specimens are available in the ASTM E1820 [94]. Readers are advised to go through the extensive literature available on this subject.
b. No correlation is available in the literature to calculate the value of the η-function for the p-SPT specimen. The procedure to develop the correlation of η-function of a p-SPT specimen is presented in a subsequent chapter in this book. Such a procedure can also be used to derive η-functions for other similar specimens.

For the information of readers, some η-functions developed for different conventional fracture specimens are presented as follows:

(i) Eqs. (2.27a) and (2.27b) show the expressions of two correlations $\eta_{J,pl}^{LLD}$ and $\eta_{J,pl}^{CMOD}$ of deeply cracked single-edge notched bend (SENB) specimens [95]. However, these expressions vary widely for shallow cracked specimens:

$$\eta_{J,pl}^{LLD} = 1.9 \tag{2.27a}$$

$$\eta_{J,pl}^{CMOD} = 3.667 - 2.199(a/W) + 0.437(a/W)^2. \tag{2.27b}$$

(ii) Sumpter [96] proposed a polynomial equation of $\eta_{plastic}$ for shallow notch bend specimens using the limit load expressions given by Haigh and Richards [97] for pure bending:

$$\eta_{J,pl}^{CMOD} = 0.32 + 12(a/W) - 49.5(a/W)^2 + 99.8(a/W)^3 \quad \text{for } a/W \le 0.282 \tag{2.28a}$$

$$\eta_{J,pl}^{CMOD} = 2 \text{ for } a/W > 0.282. \tag{2.28b}$$

(iii) Based on FEA, Sorem et al. [98] proposed an expression of the η-function for an SENB specimen with $a/W > 0.05$:

$$\eta_{J,Total}^{LLD} = 2 - \left(0.3 - 0.7\frac{a}{W}\right)\left(1 - \frac{a}{W}\right) - exp\left(0.5 - 7\frac{a}{W}\right). \tag{2.29}$$

(iv) By employing numerical analysis, Kirk and Dodds [99] carried out detailed studies of SENB specimens with shallow cracks $0.05 \le (a/W) \le 0.70$

considering material strain hardening. The η-function derived through such effort is

$$\eta_{J,pl}^{CMOD} = 3.785 - 3.101(a/W) + 2.018(a/W)^2. \tag{2.30}$$

(v) Schindler and Veidt [100] suggested expressions of η-functions for the sub-size Charpy impact test of SENB specimens:

$$\eta_{J,Total}^{LLD} = 13.818\left(\frac{a}{W}\right) - 25.124\left(\frac{a}{W}\right)^2 \text{ for } 0 \le (a/W) \le 0.275 \tag{2.31a}$$

$$\eta_{J,Total}^{LLD} = 1.90 + 0.138\left(\frac{a}{W}\right) \text{ for } (a/W) > 0.275. \tag{2.31b}$$

2.7 SIGNIFICANCE OF THE *m*-FACTOR IN FRACTURE MECHANICS

The previous section dealt with the importance of the η-function in fracture mechanics. As emphasized, the η-function correlates the global fracture parameters (LLD and CMOD) with the local crack tip parameters (CTOD and J-integral). There is one more important parameter in fracture mechanics, which is conventionally known as the m-factor. The m-factor correlates the two local crack tip parameters CTOD and J-integral as follows [101, 102]:

$$J = m_{(1)}\sigma_{ys}\delta, \tag{2.32}$$

where δ is the CTOD, σ_{ys} is the material yield stress, and m denotes plastic constraint factor. Voluminous literature is available [99, 101–103] to quantify the m-factor for different fracture specimens. Early studies by Irwin [101] showed that the m-factor is equal to 1 or 2, depending on the specimen to be considered as the plane stress or plane strain condition, respectively. This theory is based on the linear elastic fracture mechanics analysis of the energy release rate. For elastic-plastic hardening materials, Shih [102] showed that, within the region dominated by the Hutchinson–Rice–Rosengren singularity [104, 105], the m-factor in eq. (2.32) depends on material deformation properties, such as the Ramberg–Osgood strain hardening coefficient n, under small-scale yielding conditions.

Kirk and Dodds [99] and Kirk and Wang [103] proposed the following equation for the *J-CTOD* correlation for a SENB specimen, which is similar to eq. (2.32), except that σ_{ys} is replaced by flow stress σ_F:

$$J = m_{(2)}\sigma_F\delta, \tag{2.33}$$

where $\sigma_F = (\sigma_{YS} + \sigma_{UTS})/2$ and σ_{UTS} is the UTS. Based on the results of 2D plane strain FEAs, it was found that the m-factor is a function of the relative crack length (a/W), in addition to the strain hardening coefficient n. The current edition of ASTM E1820 [95] provides different forms of the m-factor as a function of a/W and σ_{ys}/σ_{UTS} for deeply cracked SENB and CT specimens. Similar equations have been proposed by many researchers for other fracture specimens. Interested readers are advised to go through the associated literature. However, it may be noted that no correlation is available in the literature to calculate the m-factor of p-SPT specimens. Chapter 6 of the present book deals with this issue.

2.8 CHAPTER CLOSURE

This chapter presents a comprehensive state of the literature available on SPT specimens. It may be noted that, in the literature, there are a few unresolved issues on SPTs. Some of these are as follows.

- Calculation of yield and ultimate stresses: The most useful correlations in SPT literature are to calculate the material yield and ultimate stresses using SPT load vs. LLD data. The proposed correlations in the literature were derived primarily for ferrous materials. A methodology to develop similar correlations for any other materials needs to be addressed.
- Biaxial fracture strain: The correlation to calculate the biaxial fracture strain using the minimum thickness data during SPT is the foundation to calculate the fracture toughness of a specimen material. In general, two empirical constants are used in such correlations. It will be of interest to readers to know the dependency of these constants on, for example, the deformation properties of the SPT specimen material, friction coefficients between the specimen and punch, and thickness of SPT specimens.
- Fracture toughness: Correlations between the biaxial fracture strain and fracture toughness have been proposed primarily using the experimental data of steels. It will be of interest to readers to know the availability of any generalized methodology to develop similar correlations for any other structural materials.
- Determination of the properties of aged materials using SPTs: One of the most important applications of the SPT methodology is to determine the change in the properties of the materials of plant components under service conditions. This needs to be carried out with the availability of a limited quantity of materials. A comprehensive case study describing the entire methodology step-by-step is desirable.

REFERENCES

[1] M.P. Manahan, A.S Argon and O.K. Harling, The development of a miniaturized disk bend test for the determination of post-irradiation mechanical properties. Journal of Nuclear Materials 103 & 104 (1981) 1545–1550.

[2] F.H. Huang, M.L. Hamilton and G.L. Wire, Bend testing for miniature disks. Nuclear Technology 57 (1982) 234.
[3] J.M. Baik, J. Kameda and O. Buck, Small punch test evaluation of inter-granular embrittlement of an alloy steel. Scripta Materialia 17 (1983) 1443–1447.
[4] J.M. Baik, J. Kameda and O. Buck, Development of small punch test for Ductile-brittle transition temperature measurement temper embrittled Ni-Cr Steels, The Use of Small-Scale Specimens for Testing Irradiated Material. ASTM STP888. W. R. Corwin and G. E. Lucas, Eds., American Society for Testing and Materials. Philadelphia (1986), pp. 92–111.
[5] X. Mao and H. Takahashi, Development of a further miniaturized specimen of 3 mm diameter for TEM disk (φ 3 mm) small punch tests. Journal of Nuclear Materials 150 (1987) 42–52.
[6] X. Mao, H. Takahashi and et al., Super small punch test to estimate fracture toughness J_{IC} and its application to radiation embrittlement of 2.25Cr-1Mo steel. Materials Science and Engineering A150 (1992) 231–236.
[7] X. Mao, M. Saito and H. Takahashi, Small punch test to predict ductile fracture toughness J_{IC} and brittle fracture toughness K_{IC}. Scripta Metallurgica et Materialia 25 (1991) 2481–2485.
[8] T. Misawa, S. Nagata and et al., Fracture toughness evaluation of fusion reactor structural steels at low temperatures by small punch tests. Journal of Nuclear Materials 169 (1989) 225–232.
[9] M. Suzuki, M. Eto and et al., Evaluation of toughness degradation by small punch (SP) tests for neutron-irradiated 2.25Cr-1Mo steel. Journal of Nuclear Materials 179–181 (1991) 441–444.
[10] W. Geary and J.T. Dutton, The prediction of fracture toughness properties from 3 mm diameter punch discs. Small Punch Test Techniques, in Proceedings of ASTM STP 1329 (1998).
[11] L. Maribel, Saucedo-Muñoz, Shi Cheng Liu and et al., Correlation between J_{IC} and equivalent fracture strain determined by small-punch tests in JN1, JJ1 and JK2 austenitic stainless steels, Cryogenics 41 (2001) 713–719.
[12] Kaishu Guan, Li Hua and et al., Assessment of toughness in long term service CrMo low alloy steel by fracture toughness and small punch test. Nuclear Engineering and Design 241 (2011) 1407–1413.
[13] C. Rodríguez, E. Cardenas and et al., Fracture characterization of steels by means of the small punch test. Experimental Fracture Mechanics 53 (2013) 385–392.
[14] J.H. Bulloch, A study concerning material fracture toughness and some small punch test data for low alloy steels. Engineering Failure Analysis 11 (2004) 635–653.
[15] T. Misawa, H. Sugawara and et al., Small specimen fracture toughness test of HT-9 steel irradiated with protons. Journal of Nuclear Materials 133&134 (1985) 313–316.
[16] T. Misawa, T. Adachi and et al., Small punch tests for evaluating ductile-brittle transition behavior of irradiated ferritic steels. Journal of Nuclear Materials 150 (1987) 194–202.
[17] J.H. Bulloch, Toughness losses in low alloy steels at high temperatures: an appraisal of certain factors concerning the small punch test. International Journal of Pressure Vessels and Piping 75 (1998) 791–804.

[18] Zhao-Xi Wang, Hui-Ji Shi and et al., Small punch testing for assessing the fracture properties of the reactor vessel steel with different thickness. Nuclear Engineering and Design 238 (2008) 3186–3193.

[19] Asif Husain, D.K. Sehgal and et al., An inverse finite element procedure for the determination of constitutive tensile behavior of materials using miniature specimen. Computational Materials Science 31 (2004) 84–92.

[20] Sisheng Yang, Zheng Yang and et al., Fracture toughness estimation of ductile materials using a modified energy method of the small punch test. Journal of Materials Research 29 (2014) 1675–1680.

[21] M. Madia, S. Foletti and et al., On the applicability of the small punch test to the characterization of the 1CrMoV aged steel: Mechanical testing and numerical analysis. Engineering Failure Analysis 34 (2013) 189–203.

[22] S. Foletti, M. Madia and et al., Characterization of the behavior of a turbine rotor steel by inverse analysis on the small punch test. Procedia Engineering 10 (2011) 3628–3635.

[23] Petr Dymáček, Stanislav Seitl and et al., Influence of friction on stress and strain distributions in small punch test models. Key Engineering Materials 417–418 (2010) 561–564.

[24] Stefan Rasche and Meinhard Kuna, Improved small punch testing and parameters identification of ductile to brittle materials. International Journal of Pressure Vessels and Piping 125 (2015) 23–34.

[25] J.M. Alegre, I.I. Cuesta and P.M. Bravo, Implementation of the GTN damage model to simulate the small punch test on pre-cracked specimens. Procedia Engineering 10 (2011) 1007–1016.

[26] E.Martinez-Paneda, I.I. Cuesta and et al., Damage modeling in small punch test specimens. Theoretical and Applied Fracture Mechanics 86 (2016) 51–60.

[27] G. Partheepan, D.K. Sehgal and et al., Fracture toughness evaluation using miniature specimen test and neural network. Computational Materials Science 44 (2008) 523–530.

[28] I.I. Cuesta and J.M. Alegre, Determination of the fracture toughness by applying a structural integrity approach to pre-cracked small punch test specimens. Engineering Fracture Mechanics 78 (2011) 289–300.

[29] E. Cardenas, F.J. Belzunce and et al., Application of small punch test to determine the fracture toughness of metallic materials. Fatigue & Fracture of Engineering Materials & Structures 35 (2012) 441–450.

[30] Yifei Xu and Kaishu Guan, Evaluation of fracture toughness by notched small punch test with Weibull stress method. Materials and Design 51 (2013) 605–611.

[31] I.I. Cuesta and J.M. Alegre, Influence of biaxial pre-deformation on fracture properties using pre-notched small punch test specimens. Engineering Fracture Mechanics 131 (2014) 1–8.

[32] T.E. Garcia, C. Rodríguez and et al., Development of a new methodology for estimating the CTOD for structural steels using the small punch test. Engineering Failure Analysis 50 (2015) 88–99.

[33] E. Martinez-Paneda, T.E. Garcia and et al., Fracture toughness characterization through notched small punch test specimens. Materials Science & Engineering A 657 (2016) 422–430.

[34] J.M. Alegre, R. Lacalle and et al., Different methodologies to obtain the fracture toughness of metallic materials using pre-notched small punch test specimens. Theoretical and Applied Fracture Mechanics 86 (2016) 11–18.

[35] E. Martinez-Paneda, I.I. Cuesta and et al., Damage modeling in small punch test specimens. Theoretical and Applied Fracture Mechanics 86 (2016) 51–60.

[36] W. R. Corwin and G. E. Lucas, The Use of Small-Scale Specimens for Testing Irradiated Materials, ASTM SPECIAL TECHNICAL PUBLICATION, (1983), STP888, DOI: 10.1520/STP888-EB

[37] W.R. Corwin, F. M. Haggag, W. L. Server, Small Specimen Test Techniques Applied to Nuclear Reactor Vessel Thermal Annealing and Plant Life Extension, ASTM SPECIAL TECHNICAL PUBLICATION, (1994), STP1204, DOI: 10.1520/STP1204-EB .

[38] M. Sokolov, J. D. Landes, G. E. Lucas, Small Specimen Test Techniques: Fourth Volume, ASTM SPECIAL TECHNICAL PUBLICATION, (2002), STP 1418, DOI: 10.1520/STP1418-EB

[39] M. Sokolov, Small Specimen Test Techniques: Fifth Volume, ASTM SPECIAL TECHNICAL PUBLICATION, (2009), STP 1502, DOI: 10.1520/STP1502-EB

[40] M. Sokolov, E. Lucon, Small Specimen Test Techniques: Sixth Volume, ASTM SPECIAL TECHNICAL PUBLICATION, (2015), STP 1576, DOI: 10.1520/ STP1576-EB .

[41] CEN Workshop Agreement, CWA 15627:2006 E, Small Punch Test Method for Metallic Materials. CEN, Brussels (2006).

[42] E.N. Campitelli, P. Spätig and et al., Assessment of the constitutive properties from small ball punch test: experiment and modeling. Journal of Nuclear materials 335 (2004) 366–378.

[43] T.E. Garcia, C. Rodríguez and et al., Estimation of mechanical properties of metallic materials by means of the small punch test. Journal of Alloys and Compounds 582 (2014) 708–717.

[44] I.I. Cuesta, C. Rodríguez and et al., Analysis of different techniques for obtaining pre-cracked/notched small punch test specimens. Engineering Failure Analysis 18 (2011) 2282–2287.

[45] J.M. Alegre, I.I. Cuesta and et al., Determination of the fracture properties of metallic materials using pre-cracked small punch tests. Fatigue & Fracture of Engineering Materials & Structures 38 (2015) 104–112.

[46] E.T. Onat and R.M. Haythomthwaite, The load carrying capacity of circular plates at large deflection. Journal of Applied Mechanics 23 (1956) 49–55.

[47] R.V. Kulkarni, K.V. Mani Krishna and et al., Determination of correlation parameters for evaluation of mechanical properties by Small Punch Test and Automated Ball Indentation Test for Zr–2.5% Nb pressure tube material, Nuclear Engineering and Design 265 (2013) 1101–1112.

[48] M.R. Bayoumi and M.N. Bassim, Study of the relationship between fracture toughness (J_{IC}) and bulge ductility. International Journal of Fracture 23 (1983) 71–79.

[49] E. Fleury and J.S. Ha, Small punch tests to estimate the mechanical properties of steels for steam power plant: I. Mechanical strength. International Journal of Pressure Vessels and Piping **75**(9) (1998) 699–706.

[50] Kundan Kumar, Arun Pooleery and et al., Evaluation of ultimate tensile strength using Miniature Disk Bend Test. Journal of Nuclear Materials 461 (2015) 100–111.

[51] Igor Simonovski, Stefan Holmström and et al., Small punch tensile testing of curved specimens: Finite element analysis and experiment. International Journal of Mechanical Sciences 120 (2017) 204–213.

[52] Ming Song, Kaishu Guan and et al., Size effect criteria on the small punch test for AISI 316L austenitic stainless steel. Materials Science & Engineering A 606 (2014) 346–353.

[53] M. Bruchhausen, S. Holmström and et al., Recent developments in small punch testing: Tensile properties and DBTT. Theoretical and Applied Fracture Mechanics 86 (2016) 2–10.

[54] R. Lacalle, J. García and et al., Analysis of key factors for the interpretation of small punch test results. Fatigue & Fracture of Engineering Materials & Structures 31 (2008) 841–849.

[55] I. Peñuelas, I.I. Cuesta and et al., Inverse determination of the elastoplastic and damage parameters on small punch tests. Fatigue & Fracture of Engineering Materials & Structures 32 (2009) 872–885.

[56] I.I. Cuesta, J.M. Alegre and et al., Determination of the Gurson–Tvergaard damage model parameters for simulating small punch tests. Fatigue & Fracture of Engineering Materials & Structures 33 (2010) 703–713.

[57] I.I. Cuesta, J.M. Alegre and et al., Influence of strain state in mechanical behaviour of aluminium alloys using the Small Punch Test. Materials and Design 54 (2014) 291–294.

[58] Jan Siegl, Petr Haušild and et al., Fractographic aspects of small punch test results. Procedia Materials Science 3 (2014) 912–917.

[59] C. Rodríguez, J. García and et al., Mechanical properties characterization of heat-affected zone using the Small Punch Test. Welding Journal 88 (2010) 188–192.

[60] M.A. Contreras, C. Rodríguez and et al., Use of the small punch test to determine the ductile-to-brittle transition temperature of structural steels. Fatigue & Fracture of Engineering Materials & Structures 31 (2008) 727–737.

[61] R. Lacalle, J. García and et al., Obtención mediante el ensayo small punch de las propiedades de tracción de materiales metálicos: Anales de Mecánica de la Fractura, 2 (2009) 501–506.

[62] Y. Xu and Z. Zhao, A modified miniature disk test for determining material mechanical properties. Journal of Testing and Evaluation 23 (1995) 300–306.

[63] J.S. Cheon and I.S. Kim, Initial deformation during small punch testing. Journal of Testing and Evaluation 24 (1996) 255–262.

[64] Y. Ruan, P. Spätig and et al., Assessment of mechanical properties of the martensitic steel EUROFER97 by means of punch tests. Journal of Nuclear Materials 307–311 (2002) 236–239.

[65] J. Isselin, and T. Shoji, Yield Strength Evaluation by Small-Punch Test. Journal of Testing and Evaluation 37 (2009) 1–7.

[66] Satya Pal Singh, Sova Bhattacharya and et al., Evaluation of high temperature mechanical strength of Cr–Mo grade steel through small punch test technique. Engineering Failure Analysis 39 (2014) 207–220.

[67] I.I. Cuesta, J.M. Alegre and et al., Effect of confinement level on mechanical behaviour using the small punch test. Engineering Failure Analysis 58 (2015) 206–211.

[68] R. Hurst and K. Matocha, Where are we now with the European code of practice for small punch testing?, in Proceedings of the International Conference on Small Sample Test Techniques, Ostrava, Czech Republic (2012) 4–18.

[69] E. Altstadt, H. Ge and et al., Critical evaluation of the small punch test as a screening procedure for mechanical properties, Journal of Nuclear Materials 472 (2016) 186–195.

[70] K. Matocha, M. Filip and et al., Determination of critical temperature of brittleness Tk0 by small punch tests, in Proceedings of COMAT 2012, Plzen Czech Republic, 21st–22nd Nov 2012.

[71] J. Chakrabarty, A theory of stretch forming over hemispherical punch heads. International Journal of Mechanical Sciences 12 (1970) 315.

[72] I.I. Cuesta and J.M. Alegre, Determination of plastic collapse load of pre-cracked Small Punch Test specimens by means of response surfaces, Engineering Failure Analysis 23 (2012) 1–9.

[73] J. A. Begley, J. D. Landes, The J-integral as a fracture criterion, in Proceedings of ASTM STP 514, Philadelphia, American Society for Testing and Materials (1972) 1-23.

[74] A.L. Gurson, Continuum theory of ductile rupture by void nucleation and growth: Part I - Yield criteria and flow rules for porous ductile media. The Journal of Engineering Materials and Technology 99(1) (1977) 2–15.

[75] V. Tvergaard and A. Needleman, Analysis of the cup-cone fracture in a round tensile bar. Acta Metallurgica 32(1) (1984) 157–169.

[76] V. Tvergaard, Influence of voids on shear band instabilities under plane strain conditions. The International Journal of Fracture 17 (1981) 389–406.

[77] V. Tvergaard, On localization in ductile materials containing spherical voids, The International Journal of Fracture 18 (1982) 237–252.

[78] C.C. Chu and A. Needleman, Void nucleation effects in biaxially stretched sheets, The Journal of Engineering Materials and Technology 102 (1980) 249–256.

[79] B.K. Dutta, In house Advanced Research in Material Damage Modelling using MADAM Code, BARC News Letter- Foundation Day Special, 189 (1999) 1–21

[80] Z.L. Zhang and E. Niemi, A class of generalized mid-point algorithms for the Gurson - Tvergaard material model. International Journal for Numerical Methods in Engineering 38 (1995) 2033–2053.

[81] B.K. Dutta, S. Guin, M.K. Sahu and M.K. Samal, A phenomenological form of the q_2 parameter in the Gurson model. International Journal of Pressure Vessels and Piping 85 (2008) 199–210.

[82] K.-H. Schwalbe, I. Scheider and A. Cornec, Guidelines for applying cohesive models to the damage behaviour of engineering materials and structures. Springer Briefs in Applied Sciences and Technology, ISBN 978-3-642-29493-8 (2013).

[83] D. S. Dugdale, Yielding of steel sheets containing slits. *Journal of the Mechanics and Physics of Solids* 8(2) (1960) 100–104.

[84] G. I. Barenblatt, The mathematical theory of equilibrium cracks in brittle fracture. *Advances in Applied Mechanics* 7 (1962) 55–129.

[85] A. Hillerborg, M. Modéer, and P.-E. Petersson, Analysis of crack formation and crack growth in concrete by means of fracture mechanics and finite elements. *Cement Concrete Research* 6(6) (1976) 773–781.

[86] A. Needleman, A continuum model for void nucleation by inclusion debonding, Journal of Applied Mechanics, 54 (1987) 525-531

[87] V. Tvergaard and J. W. Hutchinson, The relation between crack growth resistance and fracture process parameters in elastic-plastic solids, *Journal of the Mechanics and Physics of Solids* 40(6) (1992) 1377–1397.

[88] H. Yuan, G. Lin and A. Cornec, Verification of a cohesive zone model for ductile fracture Journal of Engineering Materials and Technology, 118 (1996) 192–200

[89] A. Needleman, An analysis of decohesion along an imperfect interface. In Non-Linear Fracture. Springer (1990), pp. 21–40.

[90] Z.P. Bažant, Concrete fracture models: Testing and practice. Engineering Fracture Mechanics 69(2) (2002) 165–205.

[91] I. Scheider, Cohesive model for crack propagation analyses of structures with elastic–plastic material behavior Foundations and implementation. *GKSS Research Center Geesthacht, Dept. WMS* (2001).

[92] J.M. Alegre, I.I. Cuesta and H.L. Barbachano, Determination of the fracture properties of metallic materials using pre-cracked small punch tests. Fatigue & Fracture of Engineering Materials & Structures 38 (2015) 104–12.

[93] J.M. Alegre , R. Lacalle, I.I. Cuesta and J.A. Álvarez, Different methodologies to obtain the fracture properties of metallic materials using pre-notched small punch test specimens. Theoretical and Applied Fracture Mechanics 86 (2016) 11–8.

[94] T.E. García, C. Rodríguez, F.J. Belzunce and I.I. Cuesta, Development of a new methodology for estimating the CTOD of structural steels using the small punch test. Engineering Failure Analysis 50 (2015) 88–99.

[95] ASTM Standard E1820-18a, Standard Test Method for Measurement of Fracture Toughness, Annual Book of ASTM Standards, American Society for Testing and Materials (2019).

[96] J.D.G. Sumpter, J_c determination for shallow notch welded bend specimens. Fatigue & Fracture of Engineering Materials & Structures 10 (1987) 479–93.

[97] J.R. Haigh and C.E. Richards, Yield point loads and compliance functions of fracture mechanics specimens. Central Electricity Research Laboratories (1974).

[98] W.A. Sorem, R.H. Dodds and S.T. Rolfe, A comparison of the J-integral and CTOD parameters for short crack specimen testing. Elastic-Plastic Fracture Test Methods: The User's Experience (2nd Volume). ASTM International (1991).

[99] M.T. Kirk and R.H. Dodds, J and CTOD estimation equations for shallow cracks in single edge notch bend specimens. Journal of Testing Evaluation 21 (1993) 228–38.

[100] H.J. Schindler and M. Veidt, Fracture toughness evaluation from instrumented sub-size Charpy-type tests. Small Specimen Test Techniques. ASTM International (1998).

[101] G.R. Irwin, Plastic zone near a crack and fracture toughness, in Sugamore Ordinance Materials Conference. Syracuse University Research Institute (1961).

[102] C.F. Shih, Relationships between the J-integral and the crack opening displacement for stationary and extending cracks. Journal of the Mechanics and Physics of Solids 29 (1981) 305–26.

[103] M.T. Kirk and Y.-Y. Wang, Wide range CTOD estimation formulae for SE (B) specimens. Fracture Mechanics (26th Volume). ASTM International (1995).

[104] J.W. Hutchinson, Singular behaviour at the end of a tensile crack in a hardening material. Journal of the Mechanics and Physics of Solids 16 (1968) 13–31.

[105] J.R. Rice and G.F. Rosengren. Plane strain deformation near a crack tip in a power-law hardening material. Journal of the Mechanics and Physics of Solids 16 (1968) 1–12. https://doi.org/10.1016/0022-5096(68)90013-6.

3 A Methodology to Develop New Correlations to Calculate the Yield and Ultimate Stresses of Engineering Materials using Small Punch Test Data

S.R. Ghodke and B.K. Dutta

3.1 OVERVIEW

The accuracy of determining material properties by applying SPT data primarily depends on two steps: (i) the accurate acquisition of experimental data by carefully conducting experiments and (ii) the conversion of experimental load vs. punch displacement data to material properties using empirical/semi-empirical correlations. As shown in Chapter 2, a large number of empirical correlations have been reported in the literature to determine different material properties using small punch test (SPT) data. Out of these, the most useful correlations are related to the calculation of the flow properties of materials; that is, the yield stress and ultimate stress. Multiple correlations have been reported by different authors for such a purpose. These correlations have been observed to have varying predictive capabilities. Such correlations are generally derived by determining the equation of the best fit curve passing through a large number of experimental data points for a particular group of alloys. For example, the Mao and Takahashi correlations, shown

DOI: 10.1201/9781003310372-3

as follows, for yield and ultimate stresses, were derived by primarily applying the experimental data of ferrous materials [1]:

$$\sigma_{YS} = 0.36\ P_y / t_0^2 \tag{3.1a}$$

$$\sigma_{UTS} = \frac{0.13\ P_{max}}{t_0^2} - 320. \tag{3.1b}$$

Similar correlations suggested in [8] are

$$\sigma_{YS} = 0.442\ P_y / t_0^2 \tag{3.2a}$$

$$\sigma_{UTS} = \frac{0.065\ P_{max}}{t_0^2} + 268.81. \tag{3.2b}$$

Many such examples are shown in Chapter 2 of the present book. Difficulties arise when any researcher/industry professional is interested in calculating flow properties using the SPT data of a material for which no such empirical correlations are available in the literature. To develop such correlations for a new material following existing practice in the literature, voluminous SPT specimen experimental data need to be acquired for a set of alloys of that material for which a correlation needs to be developed. This methodology is not always possible to implement as it requires the availability of a set of alloys of the material of interest.

The objective of the present chapter is to demonstrate to readers a new methodology to develop such empirical correlations analytically. To demonstrate such a procedure, two candidate materials, that is, copper and titanium, are studied here. It may be noted that such correlations for copper alloys and titanium alloys are not available in the literature. For this purpose, the mechanical properties of 14 copper alloys and 15 titanium alloys are taken from the literature. The finite element analysis (FEA) of the SPT specimens is carried out to obtain the load vs. punch displacement data of each alloy separately. To develop the correlation of the yield stress of copper alloys, the yield load (P_y) calculated using the FE results of the SPT for all the copper alloys is plotted against the standard yield stresses taken from the literature. Using the equation of the best fit curve, one can determine the correlation of the yield stress of the copper alloys. Similar exercises are carried out to obtain other correlations of the yield and ultimate stresses of both materials.

It is desirable to validate the newly developed correlations against experimental data of copper and titanium alloys. For this purpose, SPT specimens of two widely used copper alloys and one titanium alloy are fabricated and tested. The experimental data are then used to calculate the yield and ultimate stresses employing the newly developed correlations. Comparing the calculated yield and ultimate stresses with the standard literature values confirms the usefulness of the correlation. As mentioned in Chapter 2, such correlations are very useful for determining changes in material properties during the service life of an engineering component when a limited amount of the aged materials is available for testing.

3.2 PART I: ANALYTICAL DEVELOPMENT OF NEW CORRELATIONS TO CALCULATE THE FLOW PROPERTIES OF COPPER AND TITANIUM ALLOYS USING SPT DATA

3.2.1 Procedure Adopted

The salient steps used to develop empirical correlations analytically to determine the yield and ultimate stresses of copper and titanium alloys using SPT data are presented below. It may be noted that these steps can also be used for any other materials of the reader's interest.

1. The material properties, that is, the modulus of elasticity (E), Poisson's ratio (ν), and stress-strain data of 14 commercial copper alloys and 15 titanium alloys, are taken from the literature [2–6].
2. A finite element model of the SPT specimen with appropriate boundary conditions and loading arrangement is developed. It is always desirable to carry out a mesh convergence study before carrying out further numerical analysis.
3. The finite element model is then analyzed for all 14 copper alloys and 15 titanium alloys separately using the material properties collected under step 1 above from the literature. The load vs. punch displacement data thus computed for all the alloys are used to calculate (a) the value of loads at an initial localized plastic strain (P_y), traditionally called the '*yield load*', which is done by adopting the procedure suggested by Mao, and (b) the value of maximum loads (P_{max}) traditionally called the '*ultimate load*'.
4. Two plots are then developed, one for copper alloys and another for titanium alloys. The X-axis of these plots is P_y/t_0^2 using the values of P_y calculated under step 3. The Y-axis is the yield stresses of the corresponding copper or titanium alloys taken from the literature. The equation of the best fit straight line of all such data points provides the constant of the correlation $\sigma_{YS} = C_y P_y / t_0^2$. This step is repeated for copper and titanium alloys separately.
5. The procedure described under step 4 is then repeated to obtain the constant of the correlations for ultimate stresses. For this purpose, the plots are developed taking the X-axis as P_{max}/t_0^2 and Y-axis as the ultimate stresses of the corresponding alloys taken from literature. The equation of the best fit straight line of all such data points provides the constants of the correlation $\alpha_{US} = C_{US}\, P_{max}/t_0^2 + C$. This step is repeated separately for copper alloys and titanium alloys.

3.2.2 Selection of Copper and Titanium Alloys and Their Material Properties

The mechanical properties of 14 copper alloys and 15 titanium alloys are taken from the literature [2–6]. Fig. 3.1 and Fig. 3.2 show the IDs of the alloys and the corresponding mechanical properties. These properties are used later in the FEA of the SPT specimens. The stress-strain curves of all the selected copper alloys are shown in Fig. 3.1. Similarly, the stress-strain curves of all the selected titanium alloys are shown in Fig. 3.2. It may be noted that the true stress-strain curves are represented by the Holloman power law $\sigma = K\ \varepsilon^n$. The values of K and n for all the copper and the titanium alloys are also shown in these figures. It may also be noted that the range covered by the yield stresses, ultimate stresses, and hardening coefficients of the 14 copper alloys and 15 titanium alloys meet the requirements of most commercially used copper and titanium alloys of general interest.

3.2.3 FEA of SPT Specimens of Copper Alloys and Titanium Alloys to Calculate the Yield and Ultimate Loads

An SPT specimen of a 3 mm diameter disk with 0.25 mm thickness is analyzed using a finite element technique. The specimen is loaded by incrementing the

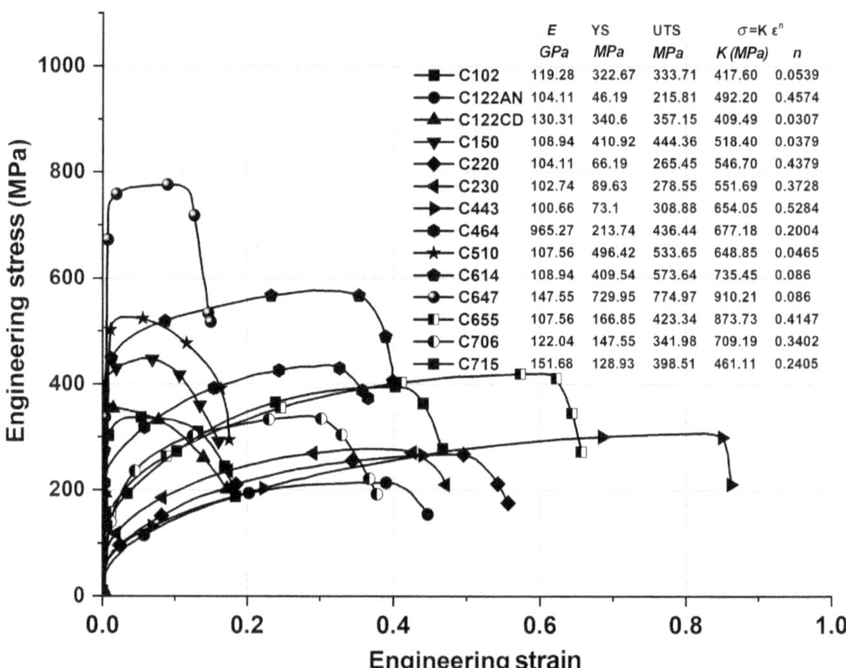

FIG. 3.1 Engineering stress-strain plots of 14 copper alloys and the corresponding mechanical properties

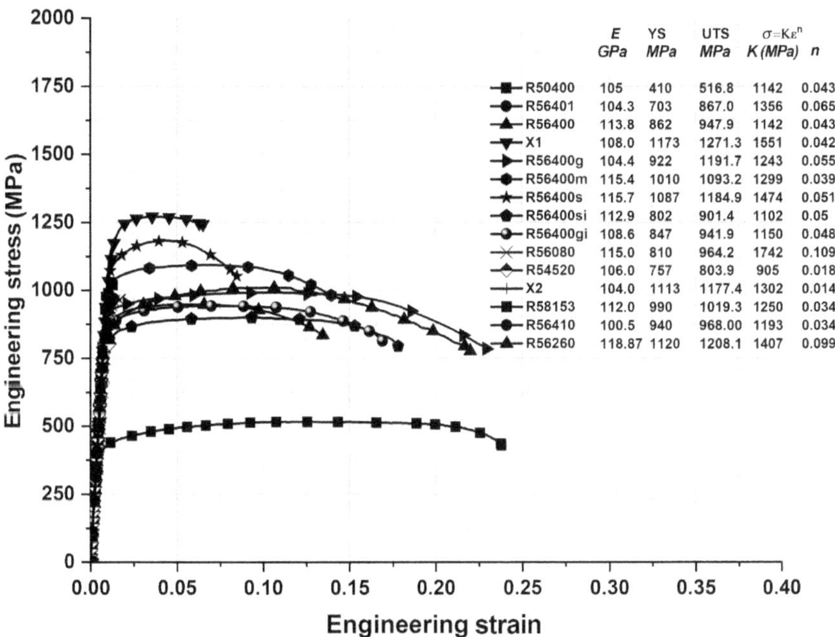

FIG. 3.2 Engineering stress-strain plots of 15 titanium alloys and the corresponding mechanical properties

displacement of the central spherical punch. The finite element mesh, appropriate boundary conditions, and loading arrangements are shown in Fig. 3.3. The analysis is carried out for all 14 copper alloys and 15 titanium alloys separately, employing the material properties shown in Fig. 3.1 and Fig. 3.2 respectively. The computed load vs. punch displacement data are shown in Fig. 3.4 and Fig. 3.5 for the copper alloys and titanium alloys, respectively.

The methodology suggested by Mao and Takahashi [1] is used here to calculate the initial localized plastic strain load P_y from all the curves shown in Fig. 3.4 and Fig. 3.5. In addition, the ultimate load P_{max} of all the curves is also obtained from the same plots. Fig. 3.4 and Fig. 3.5 also show the values of P_y, P_{max}, P_y/t_0^2, and P_{max}/t_0^2 for all 14 copper alloys and 15 titanium alloys, respectively, where t_0 is the initial thickness of the specimen in millimeters.

3.2.4 Development of Empirical Correlations to Calculate the Yield Stresses of Copper Alloys and Titanium Alloys using SPT Data

Fig. 3.6 and Fig. 3.7 show plots between the parameter P_y/t_0^2 and the yield stress of all 14 copper alloys and 15 titanium alloys, respectively. The equations of the best fit straight lines passing through all the data points are derived, and are shown below. The slopes of the straight lines are found to be 0.3586 and 0.35 for copper

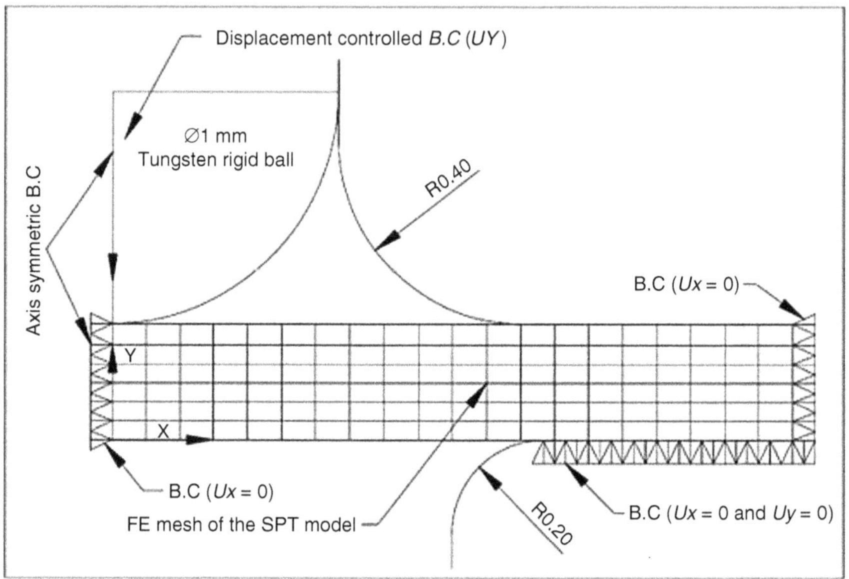

FIG. 3.3 Finite element mesh, boundary conditions, and loading arrangement of an SPT specimen

alloys and titanium alloys, respectively. These numbers are in good agreement with the suggested range between 0.3 and 0.45 in the literature for such a constant [7]. Hence, the analytically derived new correlations between σ_{YS} and P_y/t_0^2 applicable exclusively for copper alloys and titanium alloys are

$$\text{copper alloys:} \quad \sigma_{YS} = 0.3586\left(\frac{P_Y}{t_0^2}\right) \qquad \left(R^2 = 0.9417\right) \tag{3.3}$$

$$\text{titanium alloys:} \quad \sigma_{YS} = 0.35\left(\frac{P_Y}{t_0^2}\right) \qquad \left(R^2 = 0.98\right). \tag{3.4}$$

3.2.5 Development of Empirical Correlations to Calculate the Ultimate Stresses of Copper Alloys and Titanium Alloys using SPT Data

The procedure described in Section 3.2.4 is then repeated to develop empirical correlations of ultimate stresses. Fig. 3.8 and Fig. 3.9 show the plots of the parameter

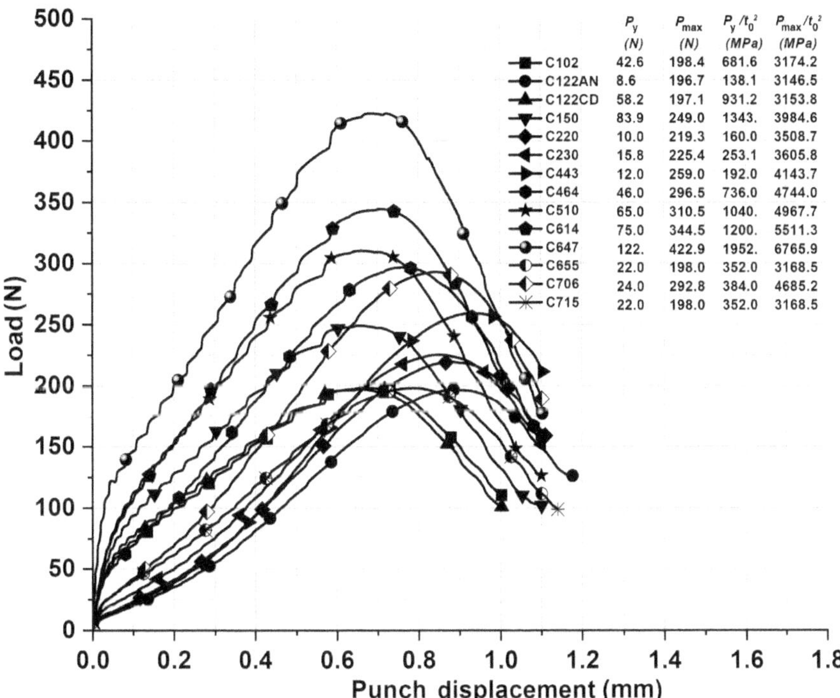

FIG. 3.4 Load vs. punch displacement plots of the SPT specimens of the 14 copper alloys calculated using FEA and the calculated values of P_y, P_{max}, P_y/t_0^2, and P_{max}/t_0^2 for each copper alloy, where P_y is the yield load, P_{max} is the ultimate load, and t_0 is the initial thickness of the SPT specimen

P_{max}/t_0^2 vs. the ultimate stress of all 14 copper alloys and 15 titanium alloys, respectively. The equations of the best fit straight lines passing through all the data points are derived, and are shown in eqs. (3.5) and (3.6) for the copper alloys and titanium alloys, respectively. Hence, analytically derived new correlations of σ_{UTS} applicable exclusively for the copper alloys and titanium alloys are

copper alloys: $$\sigma_{UTS} = 0.11315\left(\frac{P_{max}}{t_0^2}\right) - 52.435 \qquad (R^2 = 0.69) \qquad (3.5)$$

titanium alloys: $$\sigma_{UTS} = 0.11\left(\frac{P_{max}}{t_0^2}\right) + 62.06 \qquad (R^2 = 0.82). \qquad (3.6)$$

In the remaining part of the present chapter, for brevity, all the above correlations of the copper alloys will be labeled as GDCu correlations and the correlations of the titanium alloys will be labeled as GDTi correlations.

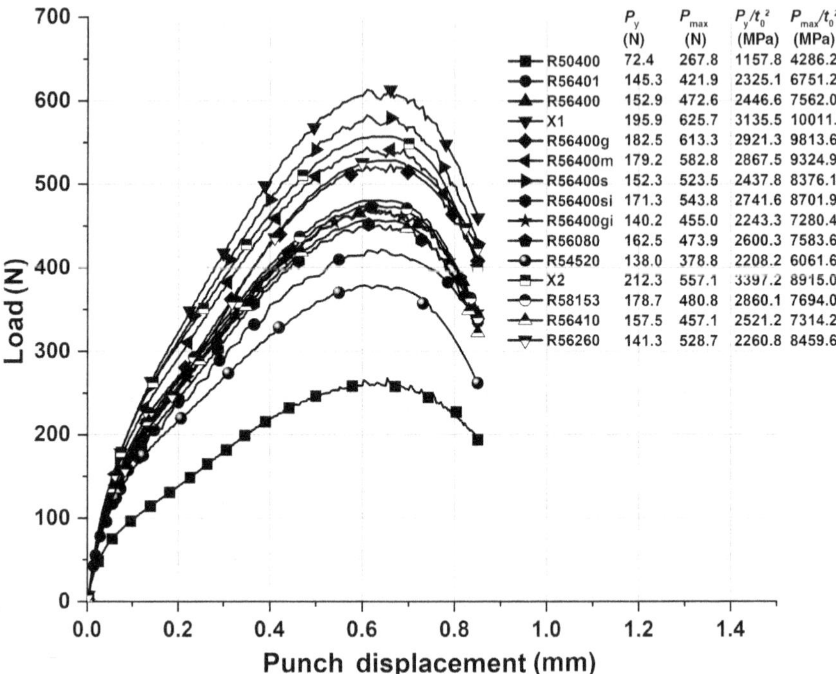

FIG. 3.5 Load vs. punch displacement plots of the SPT specimens of the 15 titanium alloys calculated using FEA and the calculated values of P_y, P_{max}, P_y/t_0^2, and P_{max}/t_0^2 for each titanium alloy, where P_y is the yield load, P_{max} is the ultimate load, and t_0 is the initial thickness of the SPT specimen

3.3 PART II: EXPERIMENTAL VERIFICATION OF THE NEWLY DEVELOPED EMPIRICAL CORRELATIONS

3.3.1 Experimental Program for SPT Specimens and the Calculation of the Yield and Ultimate Stresses using the Newly Developed Correlations

Readers are now familiar with the procedure to develop empirical correlations analytically for a group of alloys of a material of their interest to calculate the yield and ultimate stresses using SPT data. The case studies shown in Section 3.2 present the procedure used to develop such correlations for two candidate materials: copper alloys and titanium alloys. It is desirable to validate these correlations against experimental data. Such validation will not only verify the newly developed correlations but also validate the analytical procedure described above.

For this purpose, two copper alloys and one titanium alloy, that is, oxygen-free electronic (OFE) copper, copper-nickel 30% (Cu-Ni (30%)), and titanium grade 2 alloys, were selected. Multiple SPT specimens with a 3 mm diameter were fabricated using these materials. Three specimens from each material were taken

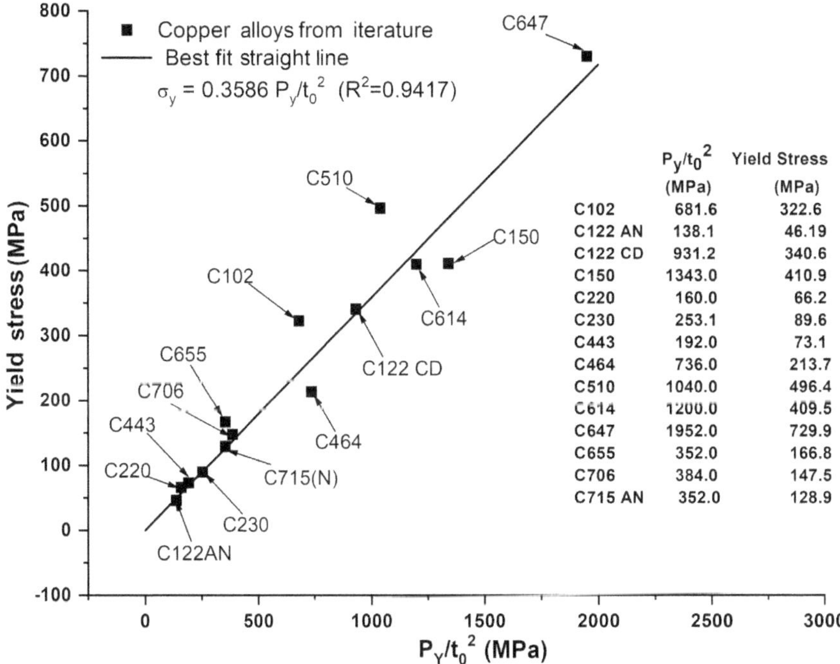

FIG. 3.6 Plot between P_y/t_0^2 calculated using the SPT data and the yield stress of the 14 copper alloys. The equation of the best fit straight line is also shown

to conduct the experiments. The specimens were tested using a conventional SPT set up.

Fig. 3.10(a) and Fig. 3.10(b) show plots of the load vs. punch displacements of all the SPT specimens of OFE copper and Cu-Ni (30%) alloy, respectively. Similar plots are also shown in Fig. 3.10(c) for the SPT specimens of the titanium grade 2 alloy. These plots are used to calculate the yield loads P_y and ultimate loads P_{max} using the procedure described in Section 3.2.3. Figs. 3.10(a–c) show the initial thickness, yield loads, and ultimate loads of all the tested specimens of OFE copper, Cu-Ni (30%) alloy, and titanium grade 2 alloy, respectively. Readers may note the minor variations in the initial thicknesses of the fabricated SPT specimens in these figures. The yield loads and ultimate loads are used as inputs to the different correlations to calculate the yield and ultimate stresses of all three alloys. The yield and ultimate stresses are calculated using (i) the newly developed correlations shown in eqs. (3.3–3.6), (ii) the correlations suggested by Mao shown in eq. (3.1a–b), and (iii) the correlations suggested by Garcia shown in eq. (3.2a–b) . These values are shown in Table 3.1 and Table 3.2 for all three alloys. Comparisons of the calculated values using these correlations and the values measured using standard tensile specimens are also presented in these tables, along with the percentage differences.

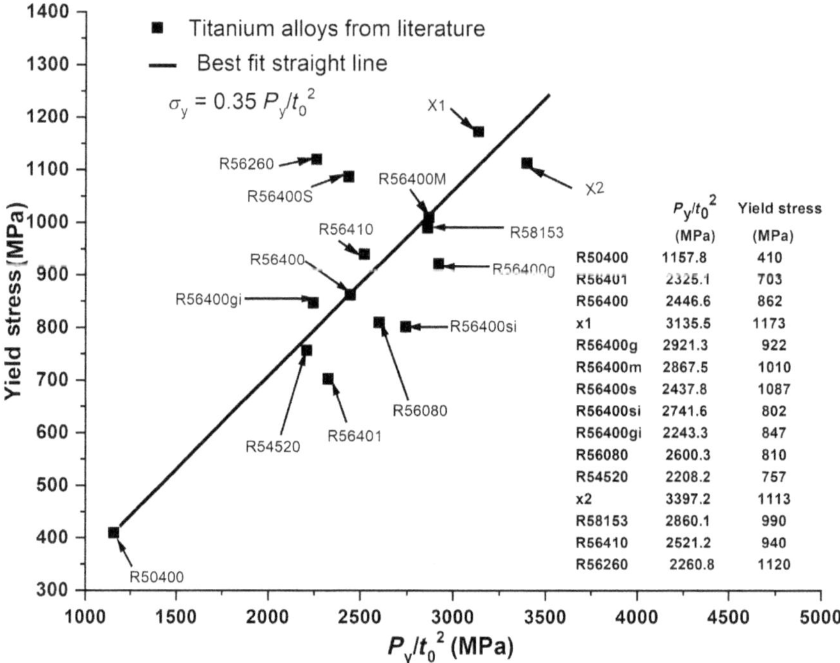

FIG. 3.7 Plot between P_y/t_0^2 calculated using the SPT data and the yield stress of the 15 titanium alloys. The equation of the best fit straight line is also shown

3.3.2 Discussion on the Calculated Yield Stresses

Readers may note from Table 3.1 that the calculated yield stress of OFE copper is 17.2% higher than the standard tensile test value when using GDCu correlation. The same value is 17.7 % higher when calculated using the Mao correlation [1] and 4.5% higher when calculated using the Garcia correlation [8]. Similar comparisons of the Cu-Ni (30%) alloy shown in Table 3.1 are -15.2% lower when calculated using GDCu correlation, -14.9% lower when calculated using Mao correlation, and 4.5% higher when calculated using Garcia correlation. Similar predictions in the case of the titanium alloy are -8.5%, -5.9%, and +15.5% when calculated using GDTi, Mao, and Garcia correlations, respectively. This leads to the conclusion that the yield stresses calculated using GDCu/GDTi correlations are close to the values calculated by the Mao correlation for all three alloys and have a good degree of consistency.

Fig. 3.11(a) shows a graphical representation of the calculated yield stresses using the three correlations. The *Y*-axis is the yield stress. Fig. 3.11(b) shows another representation of the calculated yield stresses in terms of the percentage difference with respect to the standard test values. It may be noted that, although the GDCu and GDTi correlations were exclusively developed for the respective copper

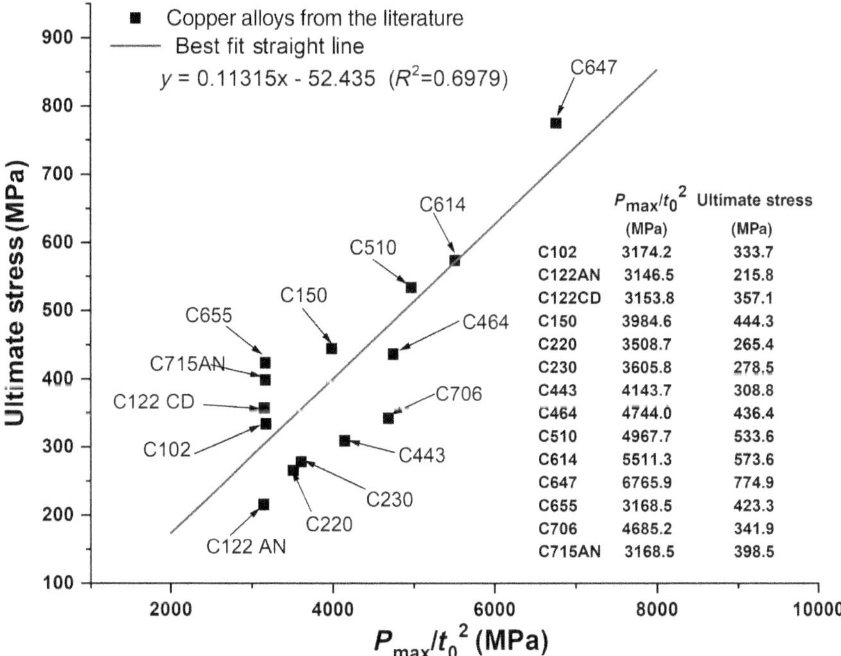

	P_{max}/t_0^2 (MPa)	Ultimate stress (MPa)
C102	3174.2	333.7
C122AN	3146.5	215.8
C122CD	3153.8	357.1
C150	3984.6	444.3
C220	3508.7	265.4
C230	3605.8	278.5
C443	4143.7	308.8
C464	4744.0	436.4
C510	4967.7	533.6
C614	5511.3	573.6
C647	6765.9	774.9
C655	3168.5	423.3
C706	4685.2	341.9
C715AN	3168.5	398.5

FIG. 3.8 Plot between P_{max}/t^2 calculated using the SPT data and ultimate stress of the 14 copper alloys. The equation of the best fit straight line is also shown

and titanium alloys, the yield stress calculated by these correlations are close to the values calculated using the Mao correlation. In addition, both correlations have a good degree of consistency in calculating the yield stress. The calculation of the yield stress using Garcia's correlation is, in general, higher than the standard test results.

3.3.3 Discussion on the Calculated Ultimate Stresses

Regarding the ultimate stress of the copper alloys, Table 3.2 shows that the calculated value using GDCu is 2.9% higher than the value quoted in the literature using the standard tensile test of OFE copper. The correlations by Mao and Garcia failed to calculate the ultimate stress of OFE copper within a desirable accuracy. A similar comparison of the Cu-Ni (30%) alloy shown in Table 3.2 is 1.2% higher when calculated using the GDCu correlation. The values calculated by the Mao and Garcia correlations are -49.2% lower and +33.5% higher than the standard tensile test results for this copper alloy, respectively. This leads to the conclusion that the GDCu correlation used to calculate the ultimate stresses of the copper alloys has the ability to calculate the values within a reasonable degree of accuracy in comparison to the standard values.

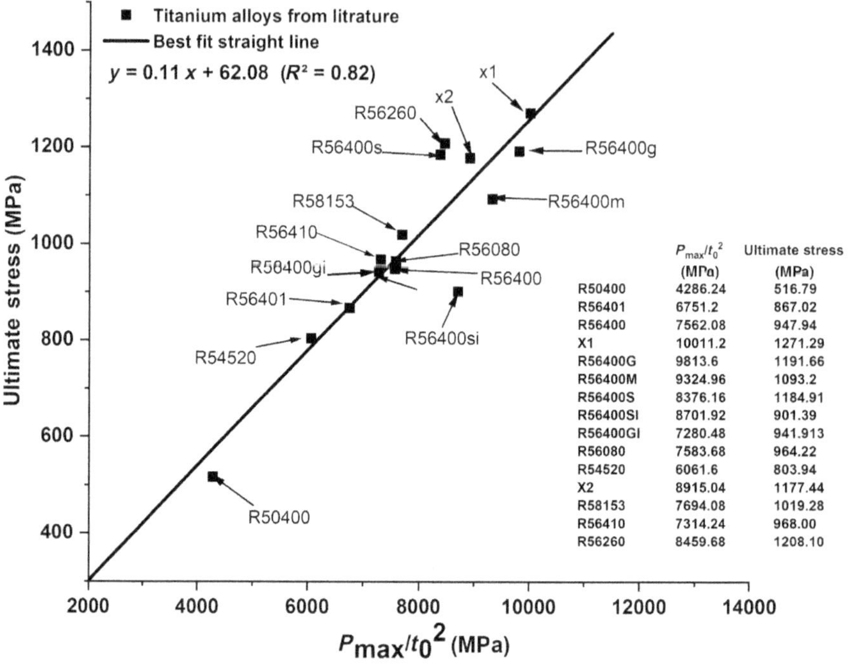

FIG. 3.9 Plot between P_{max}/t_0^2 calculated using the SPT data and ultimate stress of the 15 titanium alloys. The equation of the best fit straight line is also shown

In the case of the titanium alloy, the predicted accuracies are +17.9% higher, -32.7% lower, and 12.3% higher when using the GDTi, Mao, and Garcia correlations, respectively. These values are also shown in Table 3.2. Fig. 12(a) and Fig. 12(b) show the graphical representations of such comparisons. Fig. 12(a) shows the comparison of the absolute values of the ultimate tensile stress calculated by these three correlations and Fig. 12(b) shows the comparison of the percentage differences with respect to the values quoted in the literature using standard specimens.

3.4 CHAPTER CLOSURE

1. The purpose of the present chapter is to demonstrate to readers the procedure used to develop empirical correlations to calculate the material yield and ultimate stresses using SPT data. Such correlations are generally available in the literature for ferrous alloys employing experimental data. However, it is not always possible to obtain such experimental SPT data in literature on the material of the reader's interest. The present chapter shows an alternative method to develop such correlations analytically.
2. To demonstrate how to develop such correlations, two widely used materials, that is, copper alloys and titanium alloys, were taken. The exercise led to the

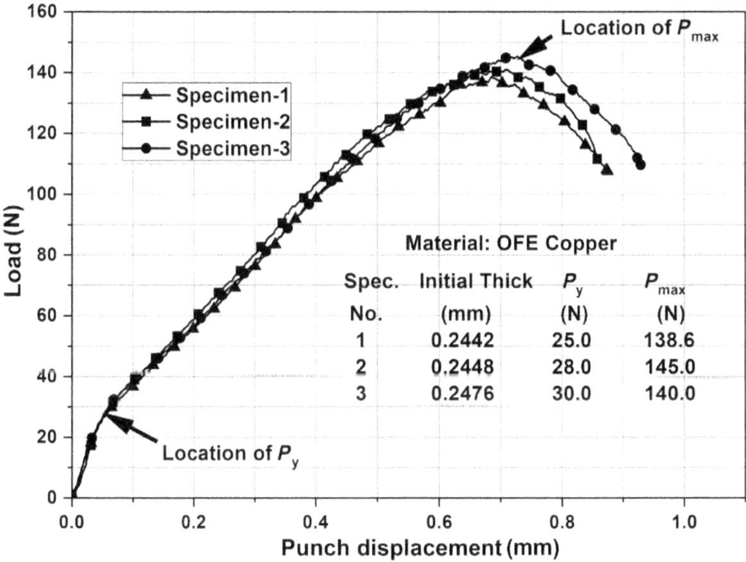

FIG. 3.10(a) Plots of the experimental load vs. punch displacement of the SPT specimens of OFE copper. The values of P_y and P_{max} for each specimen are also shown

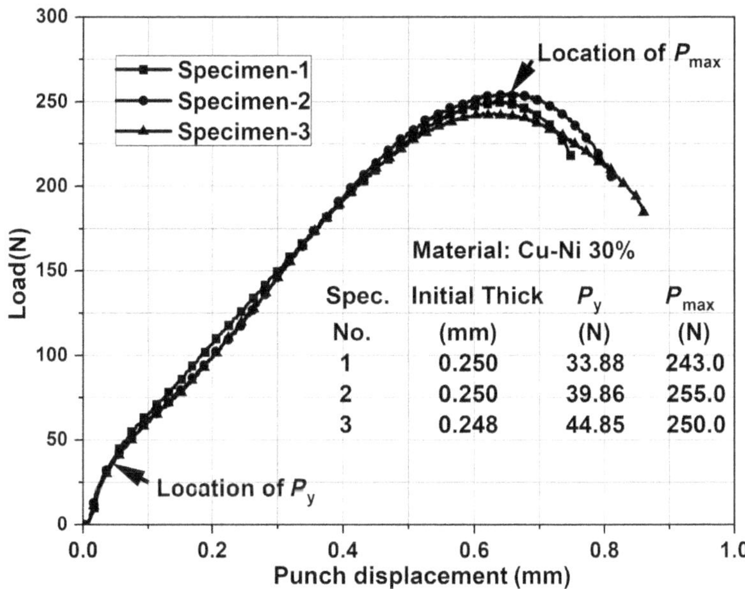

FIG. 3.10(b) Plots of the experimental load vs. punch displacement of the SPT specimens of Cu-Ni (30%). The values of P_y and P_{max} for each specimen are also shown

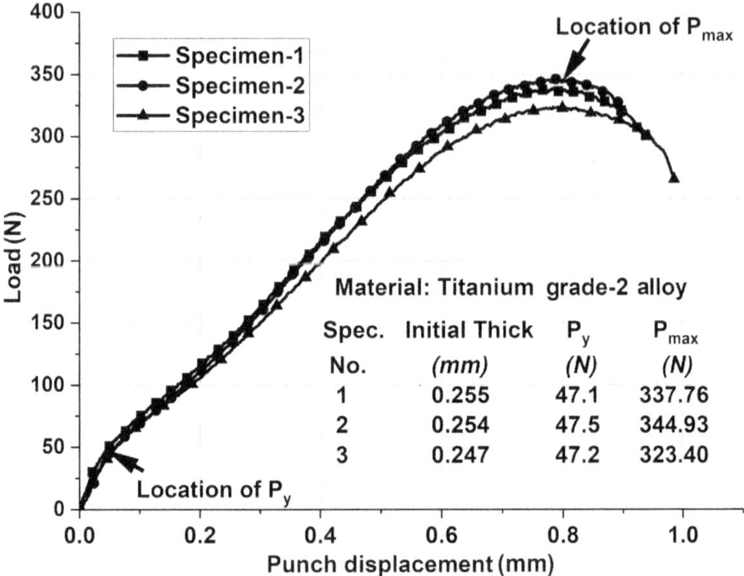

FIG. 3.10(c) Plots of the experimental load vs. punch displacement of the SPT specimens of the titanium grade 2 alloy. The values of P_y and P_{max} for each specimen are also shown

TABLE 3.1
Comparison of the yield stresses calculated using the SPT data of OFE copper, Cu-Ni (30%), and titanium grade 2 materials with tensile test data

			Yield Stress (MPa)			
	Specimen No.	P/t_0^2	GDCu	Mao	Garcia	Standard Tensile Test
OFE Cu	Sp-1	419.2	150.3	150.9	185.3	140.3
	Sp-2	467.2	167.5	168.2	206.5	
	Sp-3	489.3	175.5	176.2	216.3	
	Average	458.60	164.4	165.1	202.7	
			(+17.2%)	(+17.7%)	(+44.5%)	
Cu-Ni (30%)	Sp-1	542.1	194.4	195.1	239.6	269.2
	Sp-2	637.8	228.7	229.6	281.9	
	Sp-3	729.2	261.5	262.5	322.3	
	Average	636.4	228.2	229.1	281.3	
			(-15.2%)	(-14.9%)	(+4.5%)	
Ti-Grade 2	Sp-1	724.3	253.5	260.8	320.1	285.0
	Sp-2	736.2	257.7	265.0	325.4	
	Sp-3	773.6	270.8	278.5	341.9	
	Average	744.7	260.6	268.1	329.1	
			(-8.5%)	(-5.9%)	(+15.5%)	

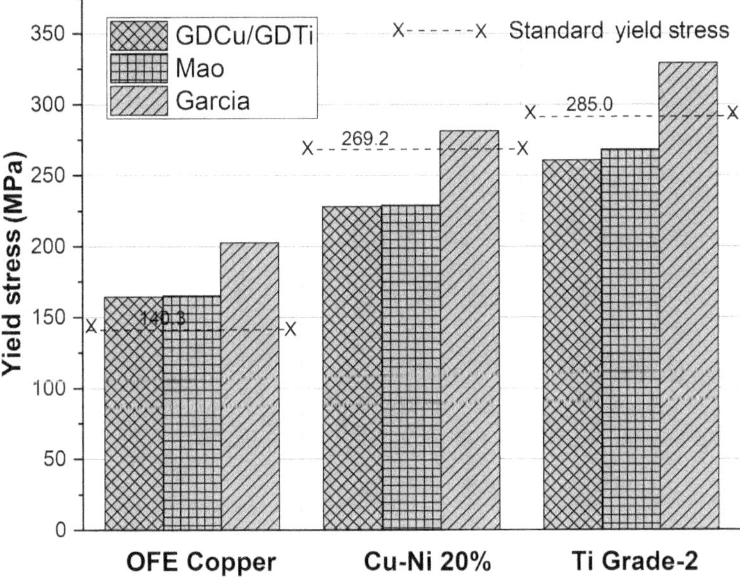

FIG. 3.11(a) Bar chart showing comparisons of the Mao, Garcia, and GDCu correlations in calculating the yield stress using experimental SPT data with respect to the standard tensile test data of the copper and titanium alloys

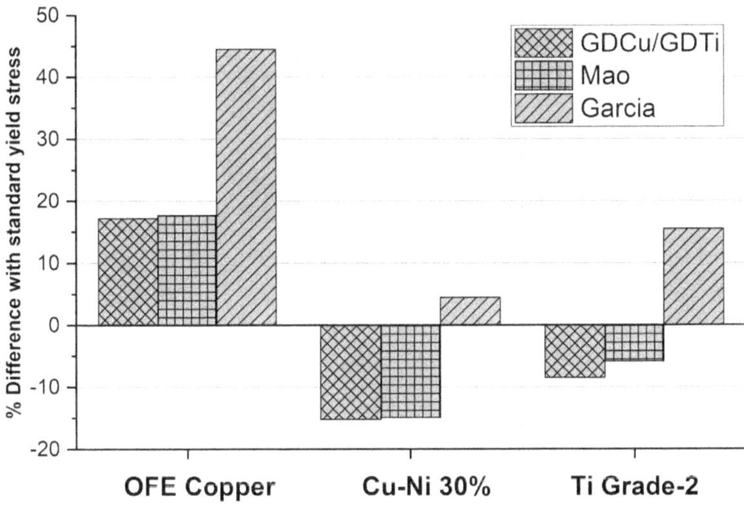

FIG. 3.11(b) Bar chart showing comparisons of the percentage difference of the yield stress calculated using the Mao, Garcia, and GDT_i correlations with respect to the standard tensile test data of the copper and titanium alloys

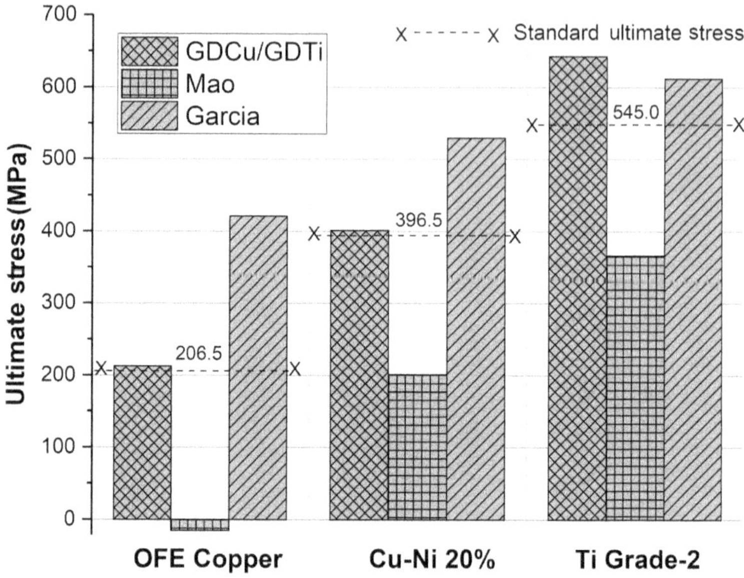

FIG. 3.12(a) Bar chart showing comparisons of the Mao, Garcia, and GDCu correlations in calculating the ultimate stress using experimental SPT data with respect to the standard tensile test data of the copper and titanium alloys

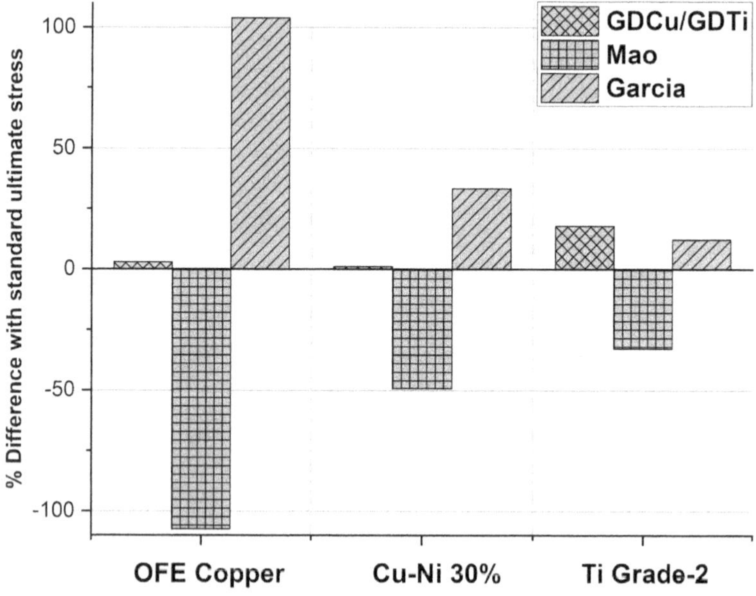

FIG. 3.12(b) Bar chart showing a comparison of the percentage difference of the Mao, Garcia, and GDT_i correlations in calculating the ultimate stress with respect to the standard tensile test data of the copper and titanium alloys

TABLE 3.2
Comparison of the ultimate stresses calculated using the SPT data of OFE copper, Cu-Ni (30%), and titanium grade 2 materials with tensile test data

			Ultimate Stress (MPa)			
	Specimen No.	P_{max}/t_0^2	**GDCu**	**Mao**	**Garcia**	**Standard Tensile Test**
OFE Cu	Sp-1	2324.2	210.5	-17.8	419.9	206.5
	Sp-2	2419.6	221.3	-5.4	426.0	
	Sp 3	2283.6	205.9	-23.1	417.2	
	Average	2342.5	212.6 (+2.9%)	-15.4 (-107.4%)	421.0 (+103.9%)	
Cu-Ni (30%)	Sp-1	3888	387.5	185.4	521.5	396.5
	Sp-2	4080	409.2	210.4	534.0	
	Sp-3	4064.8	407.5	208.4	533.0	
	Average	4010.9	401.4 (+1.2%)	201.4 (-49.2%)	529.5 (+33.5%)	
Ti grade 2	Sp-1	5194.3	633.4	355.2	606.4	545.0
	Sp-2	5346.4	650.2	375.0	616.3	
	Sp-3	5280.5	645.1	369.1	613.4	
	Average	5273.7	642.9 (+17.9%)	366.4 (-32.7%)	612.0 (+12.3%)	

development of four correlations to calculate the yield and ultimate stresses of copper alloys and titanium alloys. To develop such correlations, material stress-strain data for a group of 14 copper alloys and 15 titanium alloys were taken from the literature. Then the material stress-strain data of each alloy were used to calculate the values of P_Y/t_0^2 and P_{max}/t_0^2 by carrying out the FEA of an SPT specimen.

3. For all 14 copper alloys, a plot was then drawn between the standard yield stress quoted in the literature against the corresponding values of P_Y/t_0^2. The best fit equations derived from this plot resulted in the correlation for calculating the yield stresses of copper alloys. A similar exercise was repeated by plotting the ultimate stresses quoted in the literature against the corresponding values of P_{max}/t_0^2 for all 14 copper alloys. The best fit equations derived from this plot resulted in the correlation for calculating the ultimate stresses of copper alloys.
4. The above exercise was repeated to derive correlations to calculate the yield stress and ultimate stress of titanium alloys.
5. The second part of the present chapter was devoted to showing the significance of the newly developed correlations in comparison to existing correlations. The SPT data were obtained by conducting experiments on three widely used alloys: OFE copper, Cu-Ni (30%), and titanium grade 2. The SPT data thus

obtained were used to calculate the P_Y/t_0^2 and P_{max}/t_0^2 parameters for each material. These parameters were then used as inputs to calculate the yield and ultimate stresses by employing the Mao correlation, Garcia correlation, and newly developed GDCu/GDTi correlations. A comparison with the standard yield and ultimate stresses confirmed the importance of the GDCu and GDTi correlations.

6. It is expected that readers of this chapter will become acquainted with the analytical method to develop new empirical correlations to calculate the yield and ultimate stresses using the SPT data of a material of their interest.

REFERENCES

[1] X. Mao and H. Takahashi, "Development of a further-miniaturized specimen of 3 mm diameter for TEM disk (ø 3 mm) small punch tests," *Journal of Nuclear Materials*, vol. 150, no. 1, pp. 42–52, 1987.

[2] Application data sheet: Mechanical Properties of copper and copper alloys at low temperatures, Copper Development Association Inc., 144/8R, 1974 ".

[3] M. Frary, S. Abkowitz, S. M. Abkowitz, and D. C. Dunand, "Microstructure and mechanical properties of Ti/W and Ti–6Al–4V/W composites fabricated by powder-metallurgy," *Materials Science and Engineering: A*, vol. 344, no. 1–2, pp. 103–112, 2003.

[4] J. D. Paramore, Z. Z. Fang, M. Dunstan, P. Sun, and B. G. Butler, "Hydrogen-enabled microstructure and fatigue strength engineering of titanium alloys," *Science Reports*, vol. 7, no. 1, pp. 1–12, 2017.

[5] C. Moosbrugger, "Atlas of stress-strain curves," *ASM International, Materials Park*, 375(2002) 729–798.

[6] R. F. Muraca and J. S. Whittick, Title: Materials data handbook. Titanium 6Al-4V, p. 53, 1972.

[7] T. Misawa, S. Nagata, N. Aoki, J. Ishizaka, and Y. Hamaguchi, "Fracture toughness evaluation of fusion reactor structural steels at low temperatures by small punch tests," *Journal of Nuclear Materials*, vol. 169, pp. 225–232, 1989.

[8] T. E. García, C. Rodríguez, F. J. Belzunce, and C. Suárez, "Estimation of the mechanical properties of metallic materials by means of the small punch test," *Journal of Alloys and Compound*, vol. 582, pp. 708–717, 2014.

4 Improved Correlation between the Punch Displacement and Minimum Thickness of SPT Specimens during Tests

Pradeep Kumar and B.K. Dutta

4.1 OVERVIEW

In Chapter 2, readers became acquainted with a procedure to calculate the fracture toughness of materials using small punch test (SPT) data. One of the most important equations used to carry out such calculations is the correlation between the biaxial equivalent strain $\bar{\varepsilon}_q$ and the minimum thickness of the SPT specimen during the test. This correlation is shown as follows for ready reference:

$$\bar{\varepsilon}_q = \ln\frac{t_o}{t}, \tag{4.1}$$

where t_0 is the initial thickness of the specimen and t is the minimum thickness of the specimen at any instant during the test. As it is difficult to measure the minimum thickness (t) during the experiment, another equation is proposed in the literature [1] that correlates the minimum thickness (t/t_0) with the punch displacement (δ/t_0):

$$t/t_0 = exp\left[-\beta\left(\delta/t_0\right)^p\right]. \tag{4.2}$$

It is apparent that the punch displacement (δ) is conveniently measurable during the experiment with a great degree of accuracy in comparison to the measurement of the minimum thickness of the specimen. Eq. (4.2) involves two empirical constants, that is, β and p, which are shown to be constants in the literature. The suggested values are 2.0 and 0.9, respectively. These constants were determined by

DOI: 10.1201/9781003310372-4

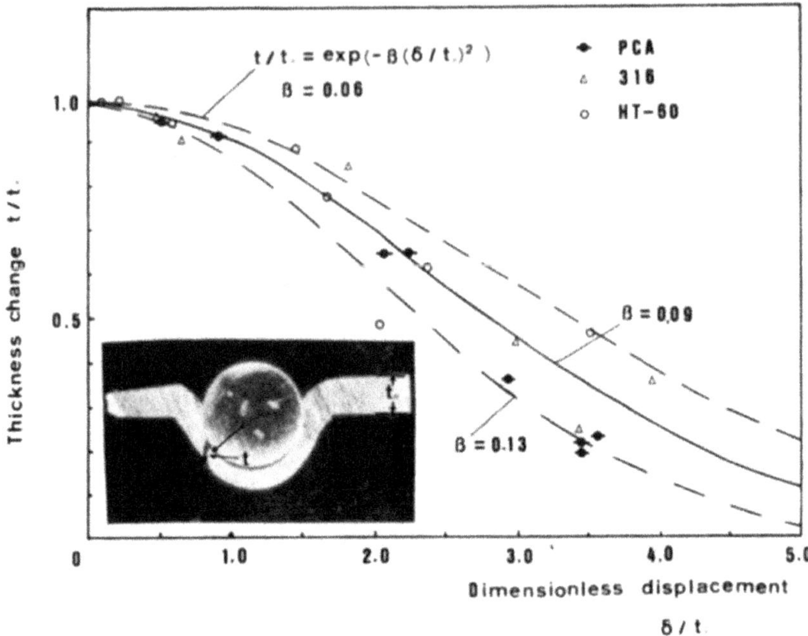

FIG. 4.1 Experimental data used to develop a new correlation between the minimum specimen thickness and punch displacement [2]

plotting a curve between t_0/t and δ/t_0 employing experimental data. For this purpose, the values of t_0/t and δ/t_0 were measured during the SPT of ferrous materials with a 0.25 mm specimen thickness. Fig. 4.1 shows such a plot. The limitations of this methodology to determine the constants β and p are as follows:

1. The empirical constants β and p are determined using the experimental data of SPT specimens of ferrous materials. It is always desirable to know the appropriateness of these constants for other materials that have significantly different properties to ferrous materials.
2. It is necessary to check the dependency of these constants on the coefficient of friction between punch and specimen surfaces.
3. It is also desirable to know the dependency of these constants on the initial thickness of the SPT specimen.

In this chapter, readers are provided with more information on the correlation between t/t_0 and δ/t_0. For this purpose, finite element (FE) analyses of the SPT specimens are carried out to examine the effect of different parameters that are expected to influence this correlation. The parameters under consideration are the material hardening coefficients, material yield stress, coefficient of friction between the punch and the SPT specimen, and the initial thickness of the SPT specimen. This exercise leads to the following conclusions.

1. The empirical constants β and p are independent of the material flow parameters, that is, the yield stress and hardening coefficients.
2. The coefficient of friction between the punch and the specimen surfaces beyond a value 0.2 also does not affect these empirical constants.
3. The initial thickness t_0 has a significant effect on these constants. Modified forms of β and p are suggested to consider the effect of the initial thickness of the SPT specimen.

Subsequently, a correlation with modified forms of the empirical constants β and p is applied to calculate the fracture toughness of unirradiated and irradiated materials using the experimental data available in the literature. This shows that the modified forms of the empirical correlations considerably improve the estimation of material fracture toughness using SPT data.

4.2 PARAMETRIC STUDIES ON SPT SPECIMENS

In the following, numerical studies are carried out to examine the effect of different parameters that are expected to influence the form of the correlation between the minimum thickness and punch displacement during the SPT. The parameters under consideration are the material hardening coefficient, material yield stress, coefficient of friction, and initial thickness of the SPT specimen.

4.2.1 FE Modeling of the SPT Specimen

A number of SPT specimens with a 3 mm diameter and varying thickness between 0.20 and 0.35 mm are modeled using the FE technique. For this purpose, a simplified axisymmetric FE model is developed for six thicknesses of the SPT specimen: 0.20, 0.225, 0.25, 0.275, 0.3, and 0.35 mm. The FE mesh is developed using eight-noded quadrilateral axisymmetric elements. A semispherical punch with a diameter of 1 mm is modeled as a rigid body. The SPT specimen is loaded under the displacement control mode by specifying incremental displacements of the punch. A schematic diagram of the FE model and boundary conditions is shown in Fig. 4.2. Readers may note that during the deformation of the SPT specimen, the top surface of the specimen has to follow the surface of the rigid spherical punch. Similarly, the bottom surface of the specimen has to follow the curved edges of the lower die. Such varying boundary conditions are accomplished using a multi-body dynamic modeling strategy. Readers are strongly advised to read about such a modeling strategy in the published literature on the FE technique. The appropriate value of the coefficient of friction is also considered between the contact surfaces of the specimen, the punch, and the lower die.

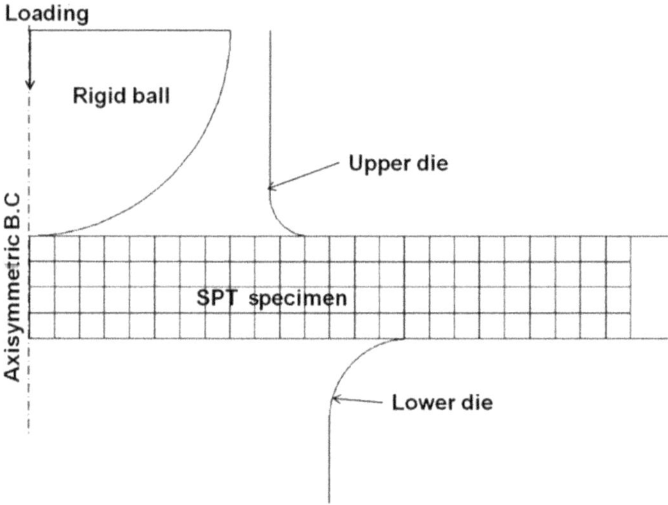

FIG. 4.2 Schematic diagram of the FE model of the SPT specimen with associated boundary conditions and loading arrangement

4.2.2 Material Modeling

Parametric analyses are carried out considering several material flow properties represented by different true stress vs. true plastic strain data. These data are generated using the Ramberg–Osgood (R-O) material model:

$$\frac{\varepsilon^p}{\varepsilon_0^e} = \alpha\left(\frac{\sigma}{\sigma_0}\right)^n, \tag{4.3}$$

where

a. ε_0^e is the elastic strain at yield, which is equal to σ_0/E,
b. ε_0^p is the plastic strain at yield, which is equal to:

$$\alpha\left(\frac{\sigma_0}{E}\right) = 0.002, \tag{4.4}$$

c. α is a dimensionless parameter,
d. n is a dimensionless material hardening parameter, and
e. ε^p and σ are the true plastic strain and true stress, respectively.

Fig. 4.3(a) shows the plots of the true stress (σ) vs. true plastic strain (ε^p) for a yield stress value of 400 MPa and material hardening coefficients of 5, 10, and 15. Similar sets of data are also generated for different values of yield stress. Fig. 4.3(b)

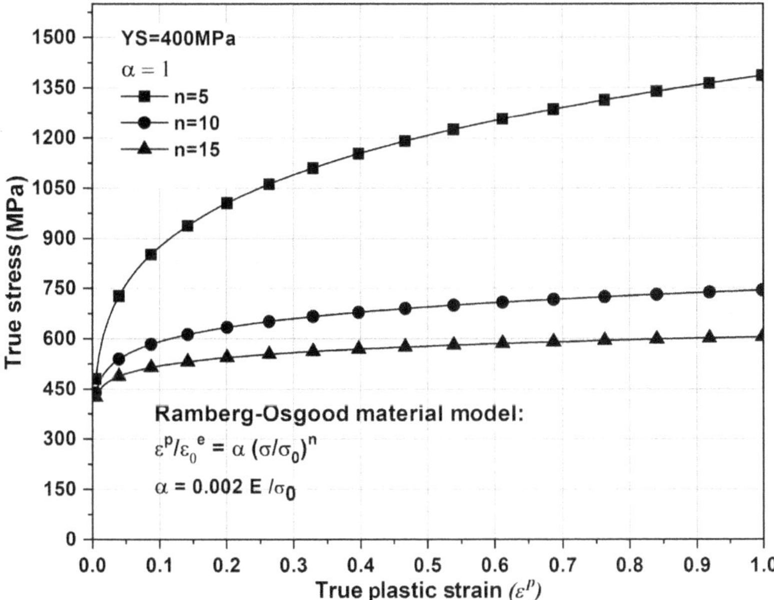

FIG. 4.3(a) Plot of true stress (σ) vs. true plastic strain (ε^p) calculated using the R-O material model for different values of the strain hardening coefficient

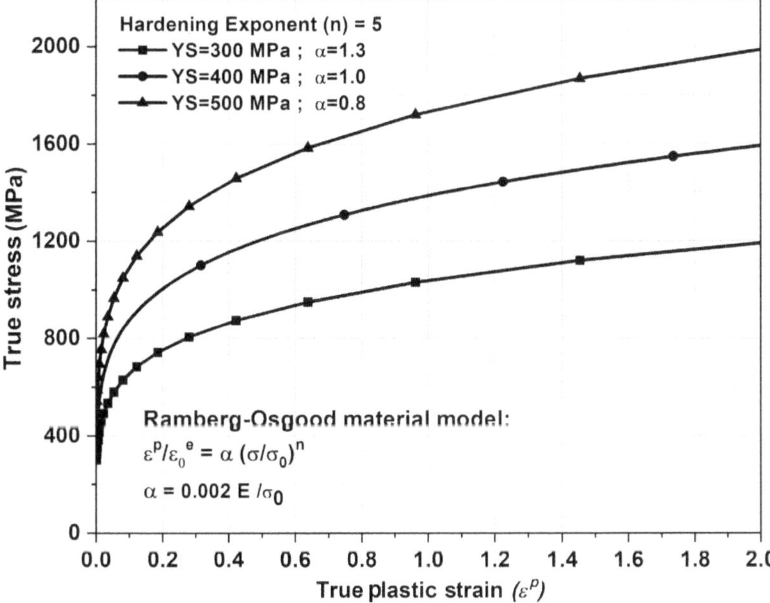

FIG. 4.3(b) Plot of true stress (σ) vs. true plastic strain (ε^p) calculated using the R-O material model for different values of yield stress

shows such plots for three yield stresses, that is, 300, 400, and 500 MPa, and a material hardening coefficient of 5. It should be noted that the R-O parameter α is a function of σ_0. Hence, different yield stresses in Fig. 4.3(b) also lead to different values of α. The corresponding values of α are also shown in this figure.

4.2.3 A Study on Parametric Variations of the Material Hardening Coefficient

The true stress vs. true plastic strain data shown in Fig. 4.3(a) for three material hardening coefficients are considered to analyze the FE model of the SPT specimen. Fig. 4.4(a) shows load vs. punch displacement plots for these analyses. In these plots, readers may note typical deformation characteristics of an SPT specimen depicting elastic bending, plastic bending, plastic membrane stretching, and plastic instability. Fig. 4.4(b) shows the plots between the minimum thickness vs. punch displacement for the same set of parameters. It may be noted that, although the load vs. punch displacement plots in Fig. 4.4(a) significantly vary with the change in material hardening coefficients, the minimum thickness vs. punch displacement plots in Fig. 4.4(b) are independent of the material hardening coefficients. This reaffirms that the correlation between the minimum thickness vs. punch displacements is independent of the material hardening coefficient, as also suggested in [2].

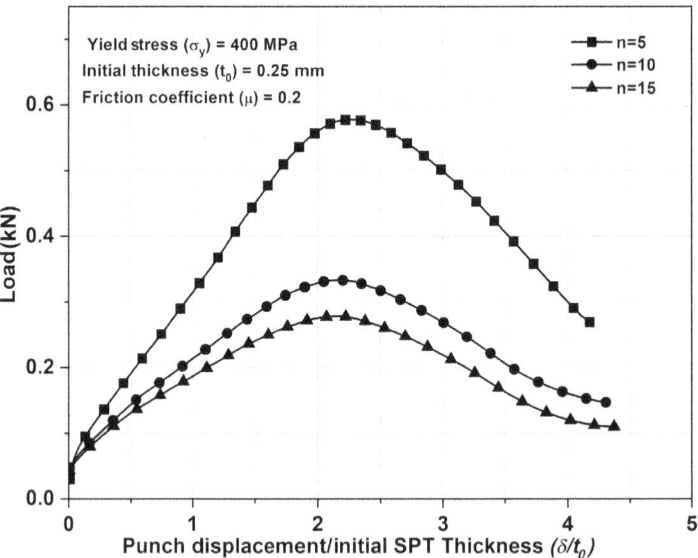

FIG. 4.4(a) Plot of the load vs. punch displacement for different values of material hardening coefficients during SPTs

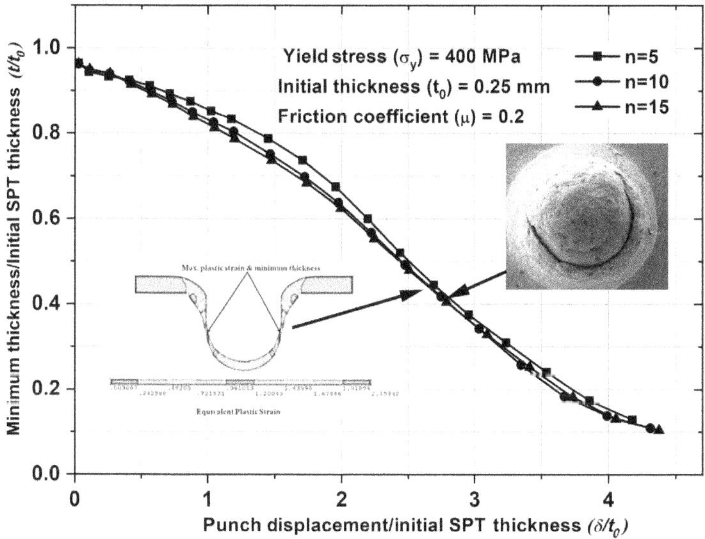

FIG. 4.4(b) Plot of the minimum thickness vs. punch displacement for different values of material hardening coefficients during SPTs

4.2.4 A Study on Parametric Variations of the Yield Stress

The parametric study is continued for different values of the yield stress. Different yield stresses also change the corresponding α value, as described in Section 4.2.2. Three values of the yield stress, that is, 300, 400, and 500 MPa, are considered for the present study. Fig. 4.5(a) shows the plots between the load and punch displacements and Fig. 4.5(b) shows the plots between the minimum thickness and punch displacements. As expected, the load vs. punch displacement plots vary significantly with the change in the yield stress. However, readers may note from Fig. 4.5(b) that the minimum thickness vs. punch displacement plots are independent of the yield stress. This again reconfirms that the correlation between the minimum thickness and punch displacements is independent of the material yield stress, as also suggested in [2].

4.2.5 A Study on Parametric Variations of the Coefficient of Friction

The coefficient of friction (μ) between the SPT specimen and the spherical punch surfaces has a significant effect on the number of output parameters. Some of these are the load vs. punch displacement, the location of the minimum thickness during the experiment, and the relationship between the minimum thickness vs. punch displacement of the specimen. FE analyses are carried out employing the model shown in Fig. 4.2. A parametric study is conducted by varying μ between the spherical punch surface and corresponding top surface of the specimen. μ is varied

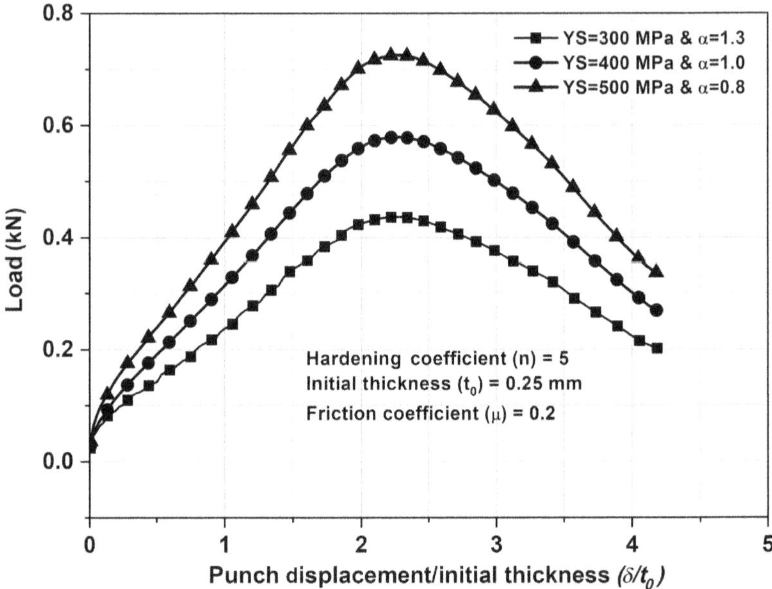

FIG. 4.5(a) Plot of the load vs. punch displacement for different values of the yield stress during SPTs

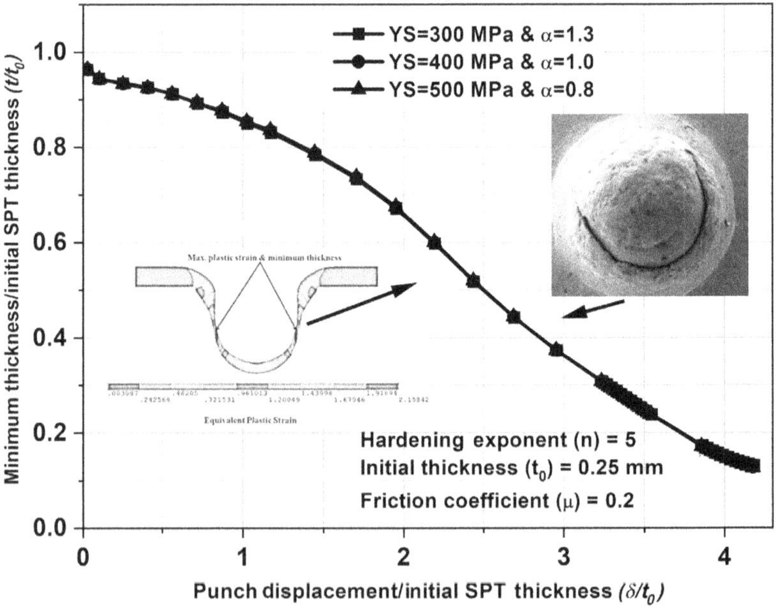

FIG. 4.5(b) Plot of the minimum thickness vs. punch displacement for different values of the yield stress during SPTs

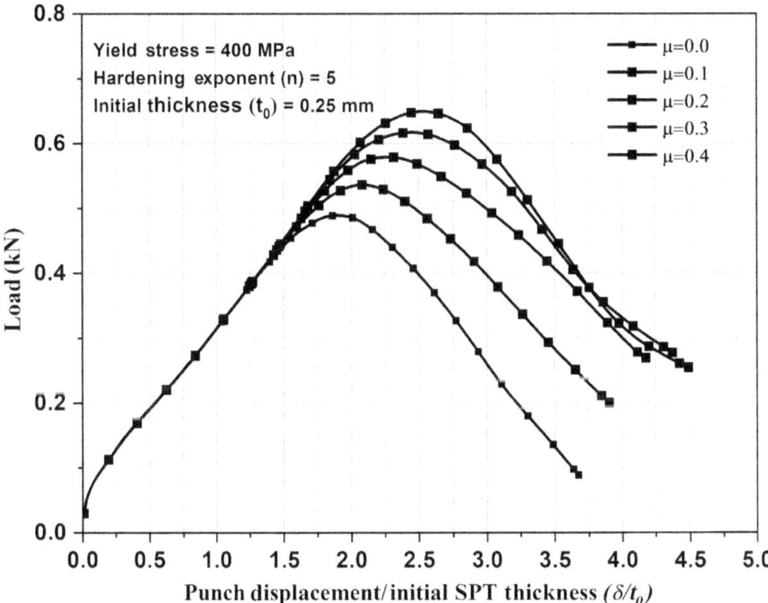

FIG. 4.6(a) Plot of the load vs. punch displacement for different values of coefficients of friction during SPTs

over the range $0.0 \leq \mu \leq 0.4$. The effect of μ on the load vs. punch displacement is shown in Fig. 4.6(a). Readers may note from Fig. 4.6(a) that the peak load increases substantially as the value of μ increases.

Fig. 4.6(b) shows the influence of μ on the minimum thickness (t/t_0) vs. punch displacement (δ/t_0). It may be noted that this correlation is a function of μ over the range $0.0 \leq \mu \leq 0.2$. However, the correlation tends to be independent of μ beyond a value of 0.2. This shows that the correlation between the minimum thickness and punch displacement is independent of the coefficient of friction only for a value of $\mu \geq 0.2$. This condition is also applicable to the correlation suggested in [2].

The location of the minimum thickness of the SPT specimen during the experiment also depends on the coefficient of friction. The present analysis shows that, for a value of $\mu < 0.2$, the location of the minimum thickness is just below the spherical punch, as shown in Fig. 4.6(b). However, the location of the minimum thickness shifts toward the outer periphery of the spherical punch for $\mu \geq 0.2$, as shown in Fig. 4.7(b). The minimum thickness location also indicates the location of the appearance of the first crack during the experiment. Hence, information on the location of the first crack during the experiment is also an indication of the order of μ.

The above observations are interpreted as follows: The correlation suggested in [2] between the minimum thickness and the center displacement is applicable for $\mu > 0.2$ as the suggested correlation is independent of μ. The scanning electron microscopy (SEM) images presented in [2] also show the circumferentially

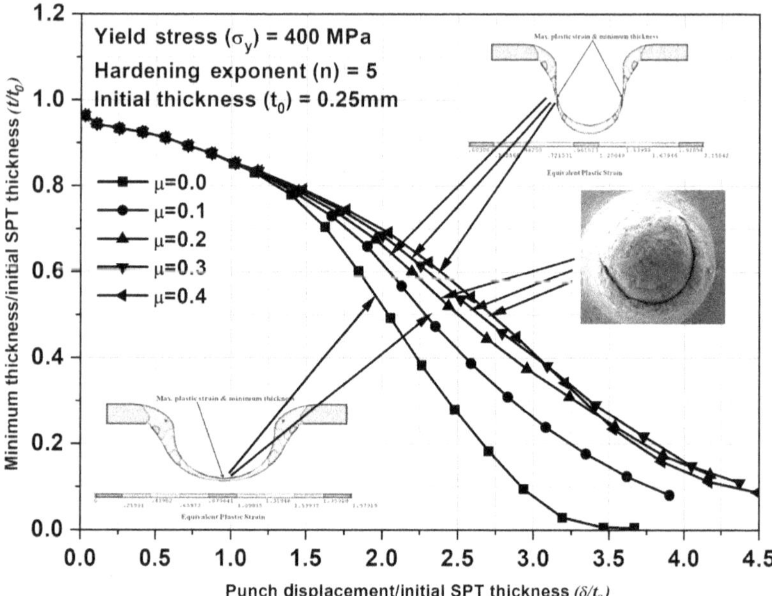

FIG. 4.6(b) Plot of the minimum thickness vs. punch displacement for different values of the coefficient of friction (μ) during SPTs. This figure also shows a shift in the location of the minimum thickness with μ

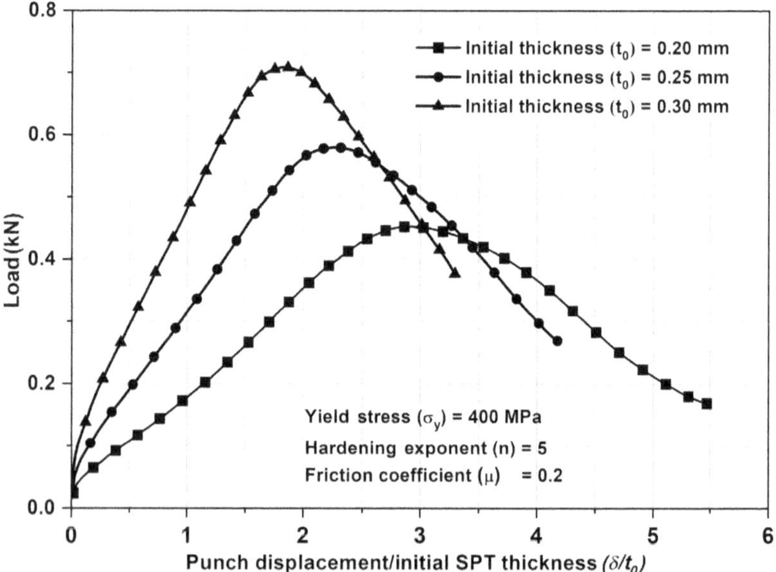

FIG. 4.7(a) Plot of the load vs. punch displacement for three values of the initial thickness of the SPT specimen

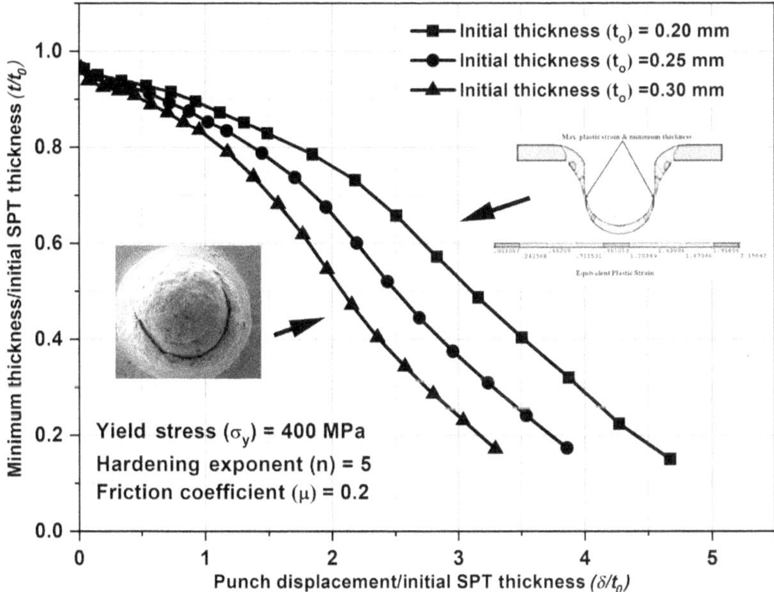

FIG. 4.7(b) Plot of the minimum thickness vs. punch displacement for three values of the initial thickness of the SPT specimen

fractured SPT specimen near the periphery of the punch. This reconfirms a value μ greater than 0.2. Hence, a correlation independent of the coefficient of friction suggested in [2] is justified. Readers may note that a circumferential crack toward the outer periphery of the punch is observed in the majority of cases during the experiment on the SPT specimen involving ductile/semi-ductile materials. Fig. 4.6(b) also shows an SEM image depicting the appearance of a circumferential crack during the experiment in the case in which the coefficient of friction $\mu > 0.2$. The SEM image was taken during an experiment conducted by the present authors. Readers should remember this limitation of eq. (4.2) while analyzing SPT data.

4.2.6 A Study on Parametric Variations of the Initial Thickness of SPT Specimens

During the experimental program for the SPT specimens, it may not always be possible to fabricate specimens with an exact thickness of 0.25 mm. It is desirable to know how the initial thickness of the fabricated specimens affects the correlation between the minimum thickness (t/t_0) and the punch displacement (δ/t_0). For this purpose, a parametric study is conducted by varying the initial thickness of the SPT specimens using the FE model, as shown in Fig. 4.2. Six initial thicknesses are considered: 0.20, 0.225, 0.25, 0.275, 0.3, and 0.35 mm. Fig. 4.7(a) shows the plots between the load vs. punch displacement and Fig. 4.7(b) shows the plots between the minimum thickness (t/t_0) vs. punch displacement (δ/t_0). For clarity, plots of only three initial thicknesses are shown. A significant effect of the initial thickness of the SPT specimen is observed in both the plots. For direct comparisons, the minimum thickness and punch displacement are normalized using the SPT initial thicknesses in Fig. 4.7(b). Readers may note that the plots remain distinctly different from each other, even after normalization, unlike in the parametric studies described earlier in Sections 4.2.3, 4.2.4, and 4.2.5. This indicates that the empirical constants β and p are functions of the initial thickness of the SPT specimens. Hence, it is necessary to derive new forms of constants β and p to take into consideration the effect of the initial thickness. Such an exercise is carried out in the next section. The new forms of β and p make the correlation between the minimum thickness (t/t_0) and the punch displacement (δ/t_0) more versatile and applicable for different values of the initial thickness of the SPT specimens.

4.3 EVALUATION OF EMPIRICAL CORRELATIONS OF β AND p AS A FUNCTION OF THE INITIAL THICKNESS OF SPT SPECIMENS

From the above four parametric studies, it is observed that the minimum thickness (t/t_0) vs. punch displacement (δ/t_0) correlation is almost independent of the material yield stress, material hardening coefficients, and a coefficient of friction greater than 0.2. However, this is not true in the case of the initial thickness of the SPT specimens. The correlation significantly depends on the initial thickness. To develop suitable correlations of β and p as a function of the initial thickness, the parametric results shown in Section 4.2.6 are used. The constants β and p are first calculated for each specimen independently assuming an exponential variation of the minimum thickness with the punch displacement, as shown in Fig. 4.8(a–c). This is carried out by employing the method of best fit exponential curve. The β and p of each specimen calculated using such a procedure are shown in Table 4.1.

To calculate β and p for any other intermediate value of the initial thickness, the best fit quadratic polynomials of β and p as a function of the initial thickness are obtained using the data shown in Table 4.1. The forms of such quadratic equations for β and p are, respectively,

$$\beta = a_1 t_0^2 + b_1 t_0 + c_1 \tag{4.5}$$

$$p = a_2 t_0^2 + b_2 t_0 + c_2. \tag{4.6}$$

The derived constants of the quadratic equations are $a_1 = 2.32$, $b_1 = 0.03$, $c_1 = -0.03$, $a_2 = 12.13$, $b_2 = -8.89$, and $c_2 = 3.38$. Fig. 4.9(a) and Fig. 4.9(b) show the plots of the quadratic forms of β and p, along with the values shown in Table 4.1.

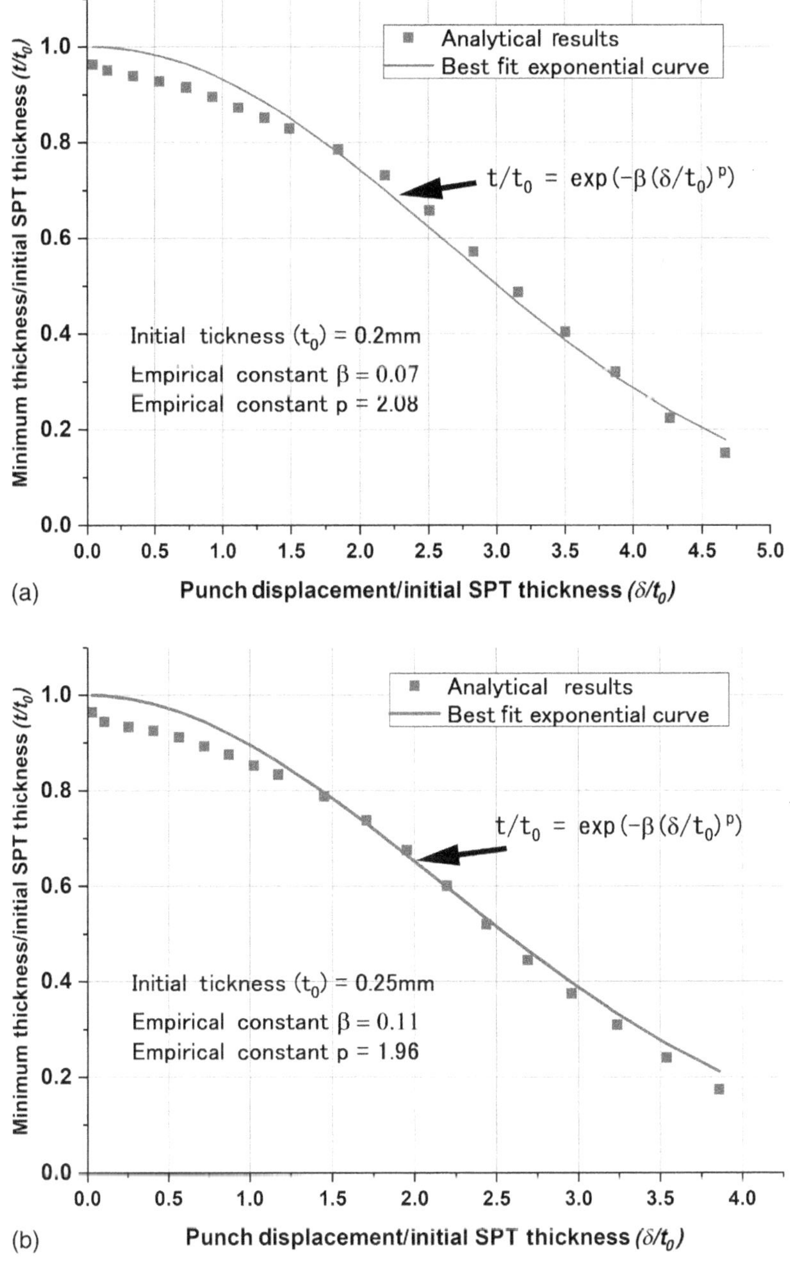

FIG. 4.8 Variation of the minimum thickness with the punch displacement for three representative SPT specimen thicknesses: (a) 0.20 mm, (b) 0.25 mm, and (c) 0.3 mm. These plots also show the empirical constants β and p of each specimen calculated independently employing best fit exponential curves

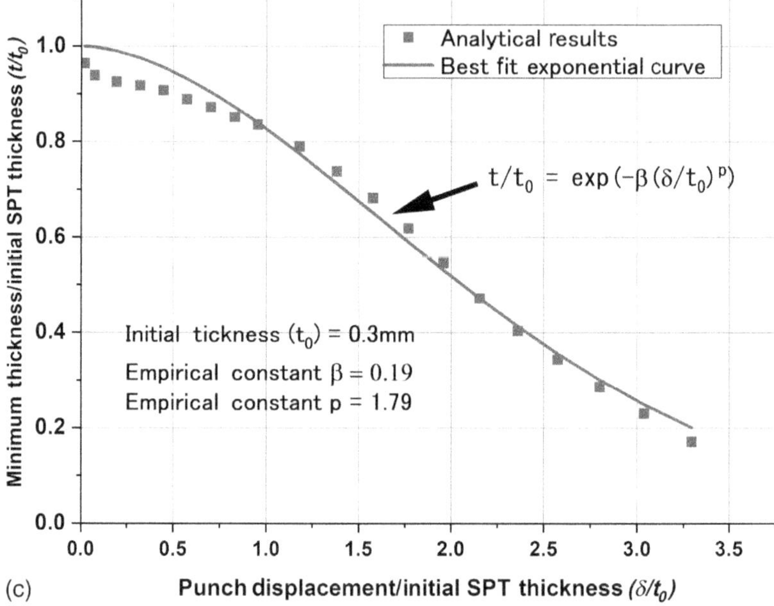

Fig. 4.8 (Continued)

TABLE 4.1
Empirical constants β and p for six initial thickness values of the SPT specimen determined by best fit curves for the FE data

Methodology	Initial thickness of the SPT specimen (t_0) (mm)	β	p
Determined by a best fit exponential curve for the FE data	0.200	0.07	2.08
	0.225	0.09	2.00
	0.250	0.11	1.96
	0.275	0.16	1.84
	0.300	0.19	1.79
	0.350	0.26	1.77
Determined experimentally [2]	≈ 0.250	0.09	2.00

4.4 CASE STUDY

A case study is presented to show the procedure of calculating the material fracture toughness using SPT data. The case study also demonstrates the importance of the quadratic forms of β and p. The experimental data presented in [3] are used for this purpose. They consist of the experimental results of SPT and conventional

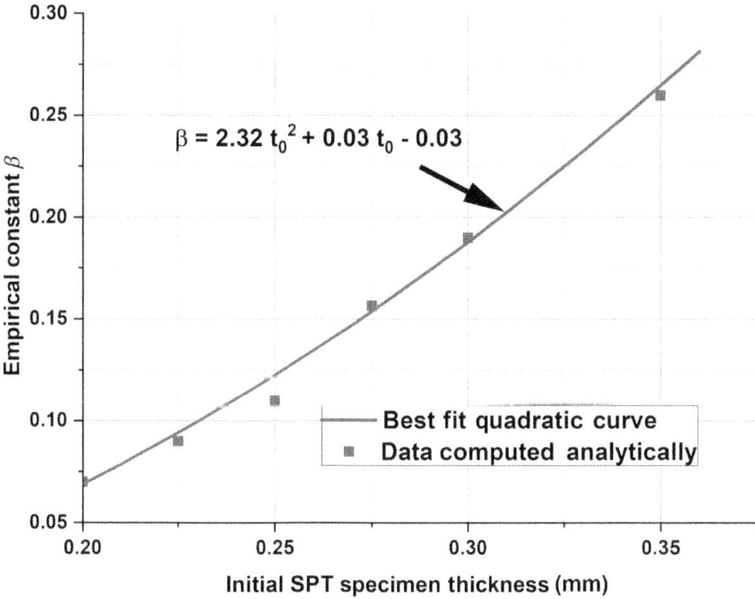

FIG. 4.9(a) Empirical constant β of the best fit quadratic curve as a function of the initial SPT specimen thickness

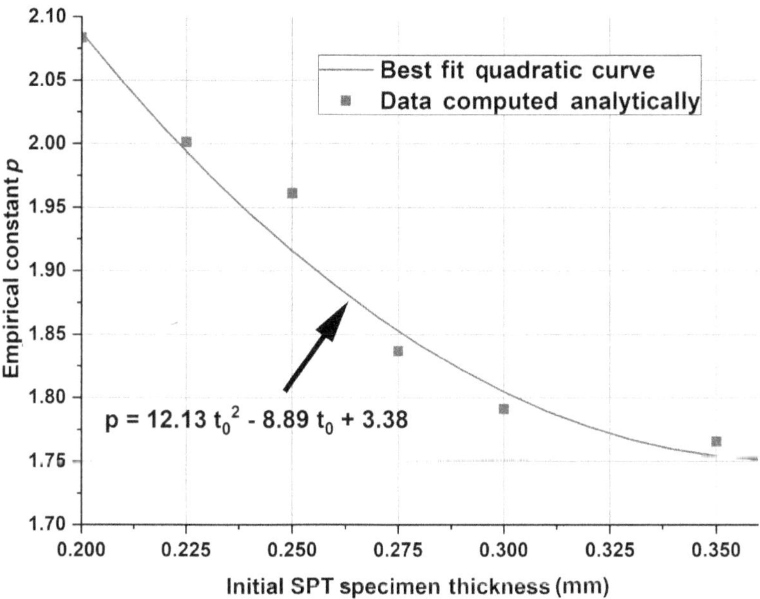

FIG. 4.9(b) Empirical constant p of the best fit quadratic curve as a function of the initial SPT specimen thickness

compact tension (CT) specimens of unirradiated and irradiated 2.25Cr-1Mo steel. A compilation of these data in tabular form for unirradiated and irradiated materials is shown in Table 4.2. The data consist of (i) the initial thickness of the specimens (t_0), (ii) the punch displacement at fracture (δ^*), (iii) the constant values of β and p suggested in [2], (iv) the calculated values of the fracture strain ε_{qf} using constant β and p, (v) the calculated fracture toughness (J_{IC}), and (vi) a comparison of the calculated fracture toughness with the experimentally measured CT data. The data shown in Table 4.2 are then recalculated using the modified forms of β and p shown in eq. (4.5) and eq. (4.6). Such modified forms of β and p are used to calculate the fracture strain ε_{qf} and subsequently, fracture toughness J_{IC}. Table 4.3 shows these revised results. It may be seen from Table 4.2 and Table 4.3 that the fracture toughness calculated by employing the quadratic forms of β and p as a function of the initial thickness is closer to the values measured using the CT specimens.

4.5 CHAPTER CLOSURE

A precise estimation of the fracture toughness using SPT data primarily depends on the accurate estimation of the biaxial fracture strain. For this purpose, it is necessary to measure the minimum thickness as accurately as possible. Such measurement is not only difficult but also vulnerable to errors. To circumvent this problem, a correlation was proposed in [2] that correlates the punch displacement with the minimum thickness during SPTs.

This correlation can be employed to calculate the minimum thickness at fracture of the SPT specimen by measuring the corresponding punch displacement at fracture. The measurement of the punch displacement is much easier than the measurement of the corresponding minimum thickness in the specimen. This correlation has two empirical constants, denoted by β and p, which are suggested to be constants in [2], and the suggested values are 0.09 and 2, respectively. The objective of the present chapter is to examine the possibility of the dependency of these two empirical constants on the material properties, coefficient of friction, and initial thickness of the SPT specimen. The outcome of the investigation reveals that the parameters β and p (i) do not depend on the material properties, that is, the yield stress and hardening coefficient, and (ii) do not depend on the coefficient of friction between the punch and specimen surfaces provided the friction factor is more than 0.2; (iii) however, they strongly depend on the initial thickness of the SPT specimens. To address the issue related to the third finding, empirical quadratic polynomials of β and p as a function of the initial thickness of the SPT specimens are derived. A case study is also presented to demonstrate the significance of employing such a correlation when calculating the fracture toughness using SPT data.

TABLE 4.2
Compilation of literature data in tabular form used in the present case study

Material 2.25Cr-1Mo steel	Initial thickness t_0 (mm)	Punch displacement at fracture δ^* (mm)	Calculated fracture strain $\bar{\varepsilon}_{qf} = \beta\,(\delta^*/t_0)^p$			J_{1C} (kJ/m^2) calculated using SPT specimen data and $J_{1C} = 345\bar{\varepsilon}_{qf} - 113$	J_{1C} (kJ/m^2) measured using the CT specimen	% difference with respect to the CT value
			β	p	$\bar{\varepsilon}_{qf}$			
Unirradiated	0.25	0.78	0.09	2	0.876	190	245	-22%
Irradiated*	0.25	0.68	0.09	2	0.666	117	148	-21%

* Temperature of irradiation: 673 K; Total Fluence: 3.2×10^{23} n/m^2

TABLE 4.3
Revised calculation of fracture toughness using the quadratic forms of β and p

Material 2.25Cr-1Mo steel	Initial thickness t_0 (mm)	Punch displacement at fracture δ^* (mm)	Calculated fracture strain $\bar{\varepsilon}_{qf} = \beta\,(\delta^*/t_0)^p$			J_{1C} (kJ/m^2) calculated using SPT specimen data and $J_{1C} = 345\bar{\varepsilon}_{qf} - 113$	J_{1C} (kJ/m^2) measured using the CT specimen	% difference with respect to the CT value
			β	p	$\bar{\varepsilon}_{qf}$			
Unirradiated	0.25	0.78	0.12	1.91	1.054	251	245	+2%
Irradiated*	0.25	0.68	0.12	1.91	0.811	167	143	+13%

REFERENCES

[1] J. Chakrabarty, A theory of stretch forming over hemispherical punch heads. International Journal of Mechanical Sciences 12 (1970) 315.

[2] X. Mao and H. Takahashi, Development of a further miniaturized specimen of 3 mm diameter for TEM disk (φ 3 mm) small punch tests. Journal of Nuclear Materials 150 (1987) 42–52.

[3] X. Mao, H. Takahashi and et al., Super-small punch test to estimate fracture toughness J_{1C} and its application to radiation embrittlement of 2.25Cr-1Mo steel. Materials Science and Engineering A150 (1992) 231–236.

5 Generalized Correlation to Ascertain Fracture Toughness using the Experimental Data of Small Punch Tests

Pradeep Kumar and B.K. Dutta

5.1 OVERVIEW

In Chapter 4, readers became acquainted with the procedure to calculate fracture toughness using small punch test (SPT) specimen data. In the following are the three primary steps used to carry out such calculations:

i) Calculate the load vs. biaxial strain data at the minimum thickness location of the SPT specimen using experimental results.
ii) Determine the biaxial strain at fracture using the data from step (i).
iii) Apply the biaxial strain at fracture to calculate fracture toughness by employing empirical/semi empirical correlations available in the literature.

In the present chapter, readers will be informed of the intricacies of the suggested correlations between the biaxial fracture strain and fracture toughness. Different empirical correlations are available in the literature. The plots of some of these correlations [1–5] are shown in Fig. 5.1. The equations of the best fit curves passing through a large set of experimental data are generally used to obtain such correlations. In Fig. 5.1, readers may observe significant differences between the correlations suggested in the literature. It may also be noted that most of the authors tried to fit a straight line between these two parameters. A linear variation implies that the fracture toughness of a material is only a function of the biaxial fracture strain. This also infers that the fracture toughness value will be unique for a group of materials that exhibit the same value of biaxial fracture strain. It may also be noted that the range of applicability of the biaxial fracture strain of the suggested correlations is limited and generally inadequate to cover all commonly

DOI: 10.1201/9781003310372-5

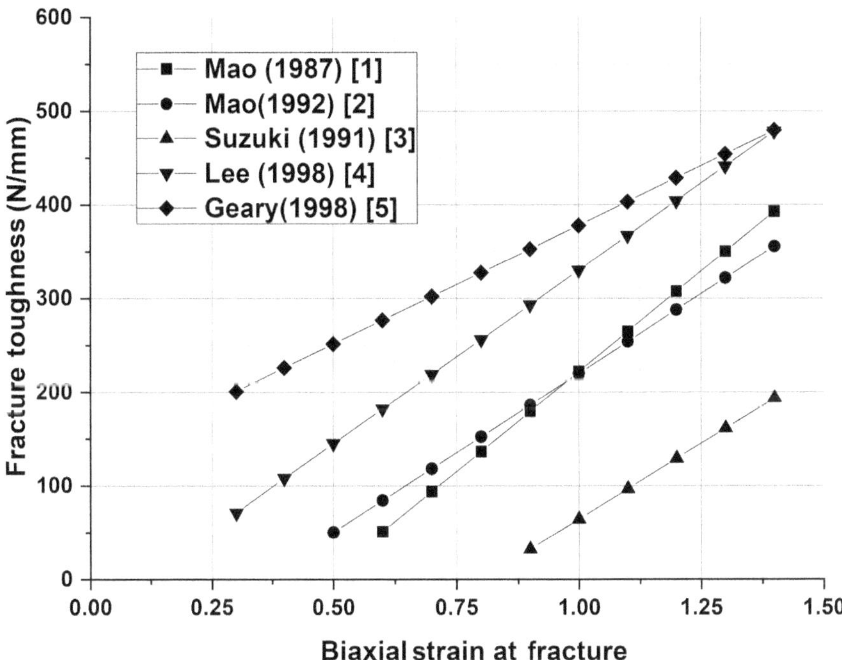

FIG. 5.1 Variations of fracture toughness with the biaxial fracture strain, as suggested by different authors in the literature

used structural steels. This is essentially due to the limited range of availability of experimental data.

The present chapter explains the development of a generalized numerical technique to develop a new correlation between the biaxial fracture strain (ε_{qf}) and fracture initiation toughness. The following issues will also be addressed in this study:

a) Is it appropriate to assume a linear correlation between the biaxial fracture strain and the material fracture toughness?
b) What are the upper bound and lower bound variations of such a correlation?
c) How can this correlation be extended to encompass most commonly used structural materials?

5.2 DETAILS OF THE NEW METHODOLOGY

To address the above issues, the present chapter deals with the development of a new correlation between the biaxial fracture strain (ε_{qf}) and fracture initiation toughness (J_i) numerically. About 100 materials are digitally generated by statistically varying the yield stress, ultimate stress, hardening coefficient, and Gurson damage parameters. It is strongly recommended that readers go through the voluminous literature available on the Gurson damage mechanics model. The aforementioned material properties can then be combined statistically to represent a group of real-life materials. Each set of aforementioned material parameters is then applied to analyze (i) an SPT specimen using the Gurson model to calculate the biaxial fracture strain ε_{qf} and (ii) a standard three-point bending (TPB) specimen to calculate fracture initiation toughness (J_i). A plot between these two calculated parameters can then be developed for all sets of the aforementioned material parameters. An equation of a best fit curve passing through all the points of this plot produces the desired correlation. This also can help to identify the upper bound and lower bound variations of the developed correlation. This procedure is explained in detail in the following sections of the present chapter.

In the remaining part of the present chapter, the newly developed correlation generated using the aforementioned procedure is evaluated against the SPT data of the four structural steels: T91, AISI403, SS304LN, and 20MnMoNi55. For this purpose, the biaxial fracture strains are calculated applying experimental data reported in [15]. The biaxial fracture strains are then used to calculate fracture initiation toughness J_i using the newly developed correlation. A comparison between the calculated J_i with the toughness values quoted in the literature using ASTM standard specimens is also presented.

5.3 STATISTICAL GENERATION OF MATERIAL DATA

As mentioned above, the properties of about 100 materials are statistically generated digitally by varying the yield stress, ultimate stress, hardening coefficient, and Gurson parameters. The variation of the yield stress is taken between 205 MPa and 580 MPa, and Ramberg–Osgood hardening exponent from 4 to 15. Each selected combination of the yield stress and hardening exponent is used to generate a complete true stress-strain curve using the Ramberg–Osgood model:

$$\frac{\varepsilon^p}{\varepsilon_0^e} = \alpha\left(\frac{\sigma}{\sigma_0}\right)^n, \tag{5.1}$$

where ε_0^e is the elastic strain at yield = σ_0/E; α is a dimensionless parameter such that $\alpha\left(\frac{\sigma_0}{E}\right) = 0.002$; n is the hardening coefficient, and ε^p and the σ are the true plastic strain and true stress, respectively. The range of yield stresses and hardening exponents considered here reasonably includes different structural steels used in real-life structures.

The Gurson parameters are also varied in a similar way. Five Gurson parameters, that is, $q_1 = 1.5$, $q_2 = 1.0$, $q_3 = 2.25$, $\varepsilon_n = 0.3$, and $S_n = 0.1$ are kept the same for all the digitally generated materials [6–8]. The initial void volume fraction f_o is assumed to be negligibly small and is equal to 10^{-5}. The remaining three Gurson parameters, that is, the void volume fraction at nucleation (f_N), void volume fraction at coalescence (f_C), and void volume fraction at failure (f_F), are varied over the ranges $0.002 \leq f_N \leq 0.03$, $0.01 \leq f_C \leq 0.15$, and $0.15 \leq f_F \leq 0.30$. It may again be noted that such ranges of Gurson parameters reasonably encompass the Gurson parameters quoted in the literature for most structural steels.

5.4 ANALYSIS OF AN SPT SPECIMEN USING THE GURSON DAMAGE MODEL

5.4.1 Specimen Dimensions and Material Parameters

An SPT specimen with a 3 mm diameter and 0.25 mm initial thickness (t_0) is modeled. A finite element (FE) model is developed using eight-noded quadrilateral axisymmetric elements. The punch is assumed to be a rigid hemispherical steel ball with a 1 mm diameter. The SPT specimen is centrally loaded with the hemispherical ball under displacement control mode. The FE model and boundary conditions are shown schematically in Fig. 5.2.

As mentioned in Section 5.3, FE analyses are carried out to calculate the load vs. central deflection for about 100 digitally generated materials with different sets of statistically varying true stress vs. true plastic strain data and the Gurson parameters.

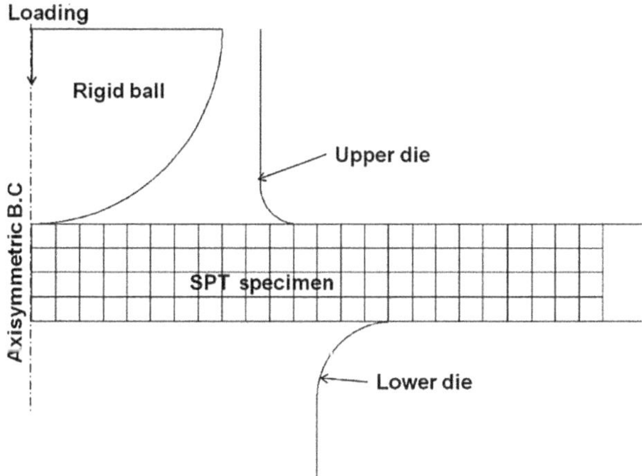

FIG. 5.2 Schematic diagram of a FE model of the SPT specimen with associated boundary conditions and loading arrangement

5.4.2 Determination of the Center Deflection at Fracture for the SPT Specimen

The change in contact surfaces during the experiment between the SPT specimen with the spherical punch and also with the lower die is simulated by invoking a multi-body dynamic modeling strategy in the FE technique. Using such a modeling strategy, the SPT specimen surface is obliged to follow the corresponding rigid surfaces. An appropriate friction factor is also considered between these surfaces.

The central deflection at fracture (δ^*) during punching is determined when the calculated void volume fractions (f) of all FEs lying along the cross-section of minimum thickness reach the material void volume fraction at fracture (f_F). Fig. 5.3 schematically shows the methodology to determine the central deflection at fracture (δ^*).

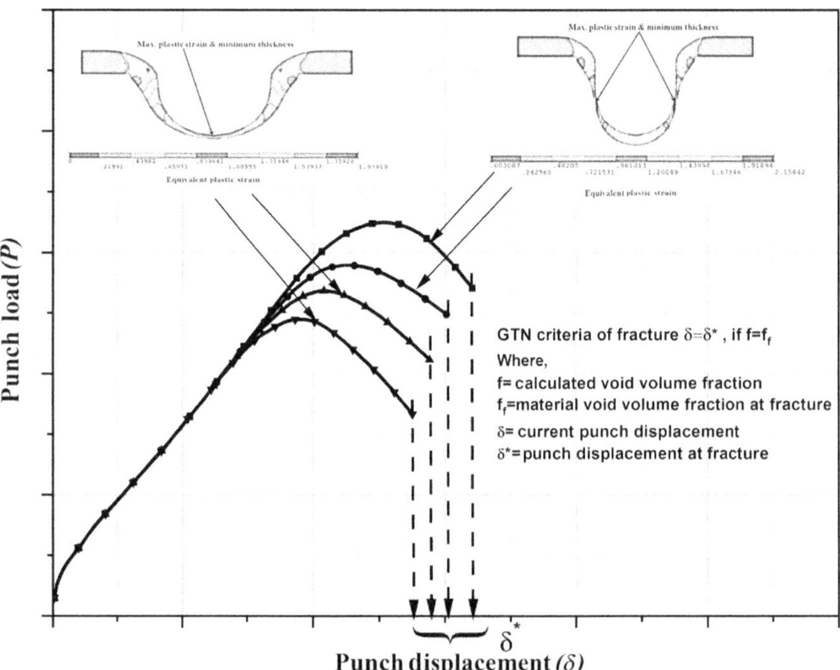

FIG. 5.3 Procedure to determine the fracture of the SPT specimen at the minimum thickness of the SPT specimen based on the void volume fraction. The location of the minimum thickness depends on the coefficient of friction between the punch and the specimen surfaces

5.4.3 Calculation of the Biaxial Fracture Strain using the FE Results

The biaxial fracture strain is then calculated using the central deflection at fracture (δ^*). The correlation used for this purpose is

$$\bar{\varepsilon}_{qf} = ln\left(t_0/t_f\right) = \beta\left(\delta^*/t_0\right)^p. \tag{5.2}$$

The empirical constants β and p are calculated using the following, as suggested in Chapter 4:

$$\beta = a_1 t_0^2 + b_1 t_0 + c_1 \tag{5.3}$$

$$p = a_2 t_0^2 + b_2 t_0 + c_2, \tag{5.4}$$

where $a_1 = 2.32$, $b_1 = 0.03$, $c_1 = -0.03$, $a_2 = 12.13$, $b_2 = -8.89$, and $c_2 = 3.38$.

The biaxial fracture strains of all the 100 digitally generated materials are determined using this methodology. It may be noted that the calculated biaxial fracture strains of the generated materials vary between 0.2 and 2.3. This range of biaxial fracture strain suffices for almost all real-life structural steels. After calculating the biaxial fracture strains using the SPT data, the next step is to calculate the fracture initiation toughness data. This is described in the next section.

5.5 DAMAGE MECHANICS ANALYSIS OF A TPB SPECIMEN APPLYING THE GURSON–TVERGAARD–NEEDLEMAN (GTN) MODEL

5.5.1 Geometrical Details and FE Model of a TPB Specimen

The ASTM E1820 standard TPB specimen is analyzed using the FE technique. The geometrical details and FE model are shown in Fig. 5.4. The eight-noded two-dimensional iso-parametric plane strain FEs are used for this purpose. Due to symmetricity, only half of the TPB specimen is modeled with appropriate symmetric boundary conditions, as shown in Fig. 5.4. The TPB model is loaded in displacement control mode at the center.

5.5.2 Evaluation of the Material's J_i and $J_{CG0.2}$ using the FE Model Results

The TPB FE model described above is then analyzed employing the GTN damage mechanics model. The geometrical variation of the q_2 parameter near the crack tip as suggested in [9] is implemented to obtain an improved assessment of J_i. The analysis of the TPB specimen is carried out by applying the properties of all the sets of digitally generated materials described above. The J-integral as a function of the central displacement is calculated as follows, as proposed in ASTM E1820:

$$J = J_{el(i)} + J_{pl(i)}, \tag{5.5}$$

where $J_{el(i)}$ is the elastic component of J, which is calculated as follows:

$$J_{el(i)} = \frac{K_{(i)}^2 (1 - \nu^2)}{E}. \tag{5.6}$$

$J_{pl(i)}$ is the plastic component of J, which is calculated using the following equation:

$$J_{pl(i)} = \left[J_{pl(i-1)} + \left(\frac{\eta_{pl(i-1)}}{b_{(i-1)}} \right) \left(\frac{A_{pl(i)} - A_{pl(i-1)}}{B_N} \right) \right] \times \left[1 - \gamma_{pl(i-1)} \left(\frac{a_{(i)} - a_{(i-1)}}{b_{(i-1)}} \right) \right], \tag{5.7}$$

where $\eta_{pl(i-1)} = 1.9$, $\gamma_{pl(i-1)} = 0.9$.

$A_{pl(i)}$ is the area under the load vs. plastic displacement at the ith load increment and b_i is the net section thickness, that is, $b_i = W - a_i$.

The crack growth as a function of the central displacement is determined by calculating the maximum distance from the crack tip of any Gauss point that has suffered complete damage. This is identified by the void volume fraction of the Gauss point becoming equal to the void volume fraction at fracture f_f.

The calculated crack growth is then plotted against the applied J-integral to obtain a complete J-R curve. A schematic plot is shown in Fig. 5.4. This figure also shows a procedure to determine the J-integral at crack initiation (J_i), as well as the J-integral at a physical crack growth of 0.2 mm ($J_{CG0.2}$).

5.6 GENERALIZED CORRELATION BETWEEN THE BIAXIAL FRACTURE STRAIN AND FRACTURE INITIATION TOUGHNESS

As mentioned above, the primary objective of the present chapter is to develop a generalized correlation between the biaxial fracture strain and fracture toughness of a material. This is done by employing the results from Section 5.4 (SPT analysis) and Section 5.5 (TPB analysis). A graph is generated between the biaxial fracture strain (ε_{qf}) calculated using the SPT specimen data and the fracture initiation toughness J_i calculated using the TPB specimen data. Fig 5.5(a) shows this plot for all the sets of digitally generated materials. Readers may observe the following points in this plot:

a) It is possible to draw a best fit straight line representing a generalized correlation between the biaxial fracture strain (ε_{qf}) and the fracture initiation toughness

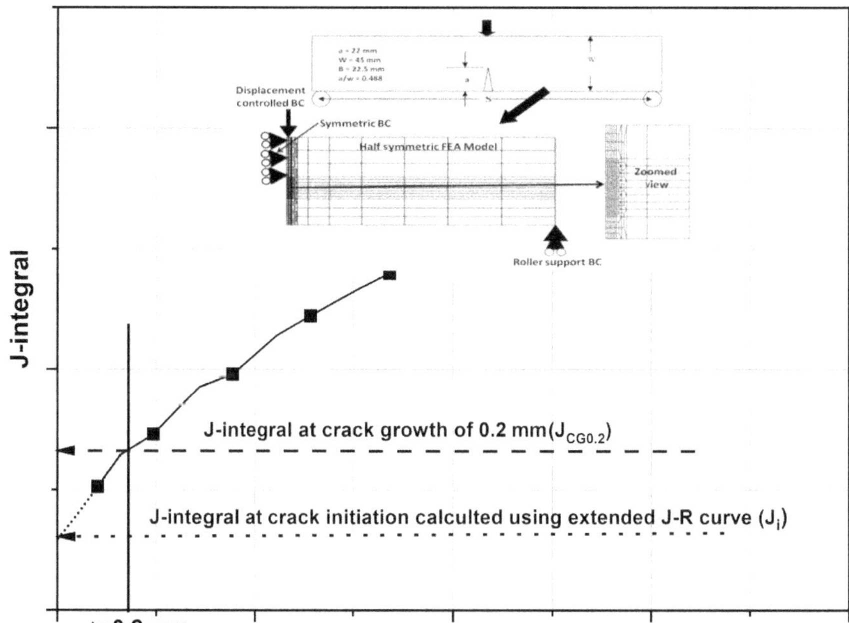

FIG. 5.4 Geometrical details and FE model of the TPB specimen and a methodology to determine J_i and $J_{CG0.2}$ from a J-R curve calculated using the FE and GTN models

(J_i). It may be noted that a number of authors have suggested a similar linear correlation between these two parameters based on the experimental data, as shown in Fig. 5.1.

b) Variations in both the lower and upper bounds of this best fit straight line are evident. These variations cannot be termed as uncertainties as each point represents a digitally generated material with definite stress-strain data and Gurson parameters. This leads to the conclusion that the biaxial fracture strain ε_{qf} does not uniquely correlate the fracture initiation toughness J_i. In other words, despite having equal biaxial fracture strains, materials can have different fracture initiation toughnesses over a lower and upper bound. Fig. 5.5(a) also describes how to quantify such a variation. The generalized correlation and the variations determined from Fig. 5.5(a) are

$$J_i = 111\bar{\varepsilon}_{qf} + 40(\pm 25)\,\text{N/mm}. \tag{5.8}$$

c) The above procedure is also repeated to obtain another generalized correlation between the biaxial fracture strain and the J-integral at 0.2 mm of physical

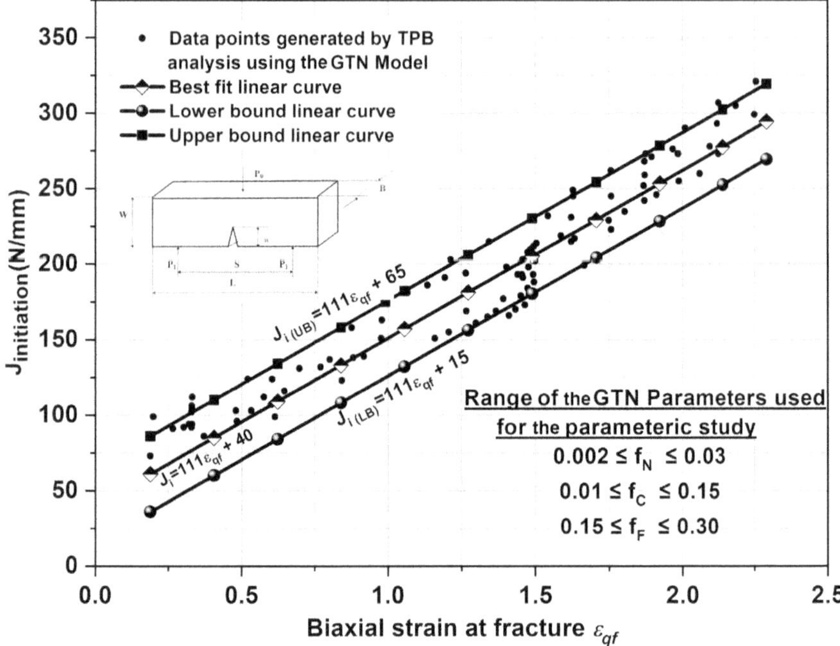

FIG. 5.5(a) New generalized correlation between fracture initiation toughness (J_i) and the biaxial fracture strain (ε_{qf})

crack growth ($J_{CG0.2}$). This is shown in Fig. 5.5(b). The correlation determined from this figure is

$$J_{CG0.2} = 143\bar{\varepsilon}_{qf} + 207(\pm 42)\,N/\text{mm}. \tag{5.9}$$

5.7 CASE STUDIES

The experimental data of SPT specimens are available in [5] for four structural steels: T91, AISI403, SS304LN, and 20MnMoNi55. Fig. 5.6 shows the plots between P/t_0^2 and biaxial strain ($\bar{\varepsilon}_q$) as measured experimentally. The biaxial strains are evaluated by applying the equations shown in eqs.(5.2)–(5.4).

The biaxial fracture strain ($\bar{\varepsilon}_{qf}$) of each material is also calculated using the procedure shown in Fig. 5.6. These are indicated by downward arrows. These values are also listed in Table 5.1. The biaxial fracture strains are characterized either by a sharp drop in the loads or the appearance of the first hunch, as shown in Fig. 5.6. The fracture initiation toughness (J_i) is then calculated using biaxial fracture strains ($\bar{\varepsilon}_{qf}$) employing eq. (5.8). The J_i data thus obtained are compared with the measured values of the four structural steels, as shown in Table 5.1. The measured values are obtained experimentally using ASTM E1820 standard

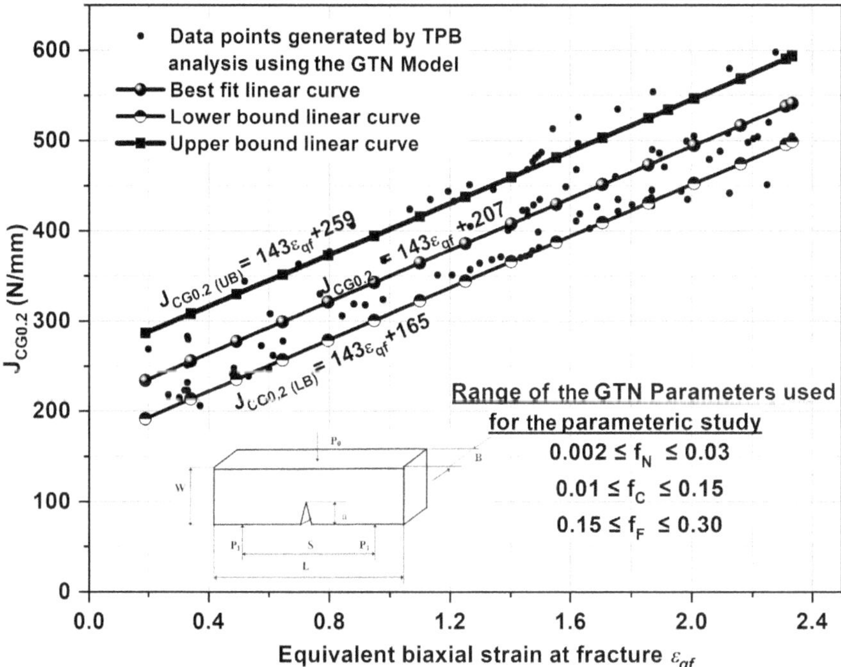

FIG. 5.5(b) New generalized correlation between the fracture toughness at 0.2 mm physical crack growth ($J_{CG0.2}$) and biaxial fracture strain (ε_{qf})

specimens and are quoted in references [10–14]. It may be seen from Table 5.1 that the predicted fracture initiation toughness (J_i) calculated by the proposed generalized correlation is comparable with the measured values of the ASTM E1820 TPB standard tests.

5.8 CHAPTER CLOSURE

In the present chapter, readers became familiarized with the calculation of fracture toughness using SPT specimen data. Such calculations are performed by applying empirical correlations between the biaxial fracture strain ($\overline{\varepsilon}_{qf}$) and fracture initiation toughness (J_{IC}). Several such correlations are available in the literature. These correlations are derived using experimental data. The observations on such correlations are as follows:

a) Most of the authors proposed a linear correlation between these two parameters.
b) The linear correlation leads to the conclusion that a set of materials exhibiting equal biaxial fracture strains also have the same fracture initiation toughness.

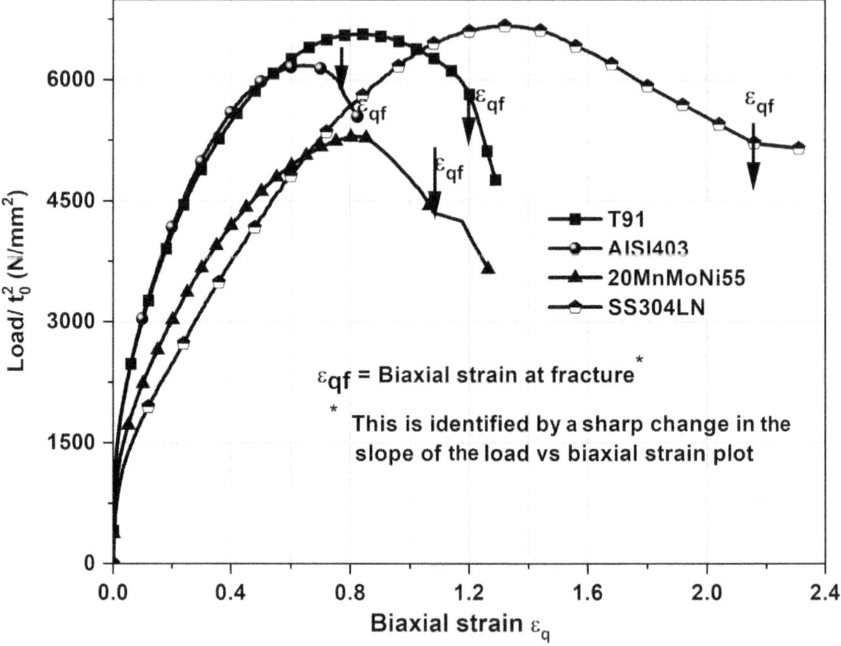

FIG. 5.6 Plots of the load vs. biaxial strain of the SPT specimens of four structural steels and identification of the biaxial fracture strain

TABLE 5.1
Comparison of the predicted J-initiation with the measured values of four structural steels

Material ID	Biaxial fracture strain, $\varepsilon_{qf} = \beta(\delta^*/t_0)^p$	Fracture initiation toughness J_i (N/mm): Predicted using proposed generalized eq. (5.8)	Fracture initiation toughness J_i (N/mm): Values reported in the literature using ASTM E1820 standard tests
T91	1.225	176 ± 25	198
AISI403	0.770	125 ± 25	100
SS304LN	2.157	279 ± 25	300
20MnMoNi55	1.083	160 ± 25	150

c) The experimental data limit the validity range of each correlation. Due to such a limitation, an individual correlation does not encompass all the commonly used structural steels in general.

To overcome these difficulties, the present chapter explained a unique method for generating such a correlation analytically. The inferences are as follows:

a) A linear correlation between these two parameters reported in the literature is justified.
b) The correlation has upper bound and lower bound variations. This means that a group of materials exhibiting equal biaxial fracture strain may have a variation in the fracture initiation toughness.
c) The present analytical exercise also helped to derive a generalized correlation that has a wider range of applicability covering most commonly used structural steels.
d) A case study was also presented to describe the ability of the newly derived correlation to predict fracture initiation toughness (J_i) in comparison to the experimental values.

REFERENCES

[1] X. Mao X, T. Shoji, H. Takahashi, Characterization of fracture behavior in Small Punch Test by Combined Recrystallization-Etch Method and Rigid Plastic Analysis, Journal of Testing and Evaluation 15 (1) (1987) 1–8, DOI: 10.1520/JTE11549J.
[2] X. Mao, H. Takahashi, T. Kodaira, Super Small Punch Test to estimate fracture toughness J_{IC} and its application to radiation embrittlement of 2.25Cr-1Mo steel, Materials Science and Engineering AI50 (1992) 231–236.
[3] M. Suzuki, M. Eto, K. Fukaya, Y. Nishiyama, T. Kodaira, M. Oku, et al., Evaluation of toughness degradation by Small Punch (SP) Tests for neutron-irradiated 2iCr-1Mo steel, Journal of Nuclear Materials (1991) 179–181, 441–444.
[4] W.K. Lee, D.R. Metzger, A. Donner, O.E. Lepik, Small specimen test techniques. ASTM, STP 1329, Philadelphia (USA). (1998) 539–556.
[5] W. Geary, J.T. Dutton, Small specimen test techniques. ASTM, STP 1329, Philadelphia (USA). (1998) 588–601.
[6] V. Tvergaard, Influence of voids on shear band instabilities under plane strain conditions, International Journal of Fracture 17 (1981) 389–406.
[7] V. Tvergaard, On localization in ductile materials containing spherical voids, International Journal of Fracture 18 (1982) 237–252.
[8] C.C. Chu, A. Needleman, Void nucleation effects in biaxially stretched sheets, Journal of Engineering Materials and Technology 102 (1980) 249–256.
[9] B.K. Dutta, S. Guin, M.K. Sahu, M.K. Samal, A phenomenological form of the q2 parameter in the Gurson model, International Journal of Pressure Vessels and Piping 85 (2008) 199–210.
[10] Contract report. Flow and fracture behavior of 9%Cr-ferrritic/martensitic steels, SCK.CEN-R-4122 (2005).

[11] C. Gupta, J.S. Dubey and et al., Characterization of fatigue and fracture behaviour and improvement of toughness of end fitting material of pressurized heavy water reactors, High Temperature Materials and Processes 16 (1) (1997) 65–75.

[12] M.I. de Vries, B. Schaap, Experimental observations of Ductile Crack Growth in type 304 stainless steel, Elastic-Plastic Fracture Test Methods: The User's Experience. ASTM STP 856, E.T. Wessel and F.J. Loss, Eds., American Society for Testing and Materials (1985) 183–195.

[13] M.S. El-Fadaly, T.A. El-Sarrage, A.M. Eleiche, W. Dahl, Fracture toughness of 20MnMoNi55 steel at different temperatures as affected by room-temperature pre-deformation, Journal of Materials Processing Technology 54 (1995) 159–165.

[14] E. Lucon, The use of single-specimen techniques for measuring upper shelf toughness properties under impact loading rates. SCK •CEN - BLG – 1016 (2005).

[15] Pradeep Kumar, B.K. Dutta, J. Chattopadhyay, R.S. Shriwastaw, Numerical evaluation of J-R curve using small punch test data, Theoretical and Applied Fracture Mechanics 86 (2016) 292–300, doi 10.1016/j.tafmec.2016.08.003.

6 Development of the η_{pl}-Function and *m*-Factor of p-SPT Specimens

Taslim D. Shikalgar, B.K. Dutta, and J. Chattopadhyay

6.1 OVERVIEW

The importance of the η_{pl}-function and m-factor in fracture mechanics is presented in Chapter 2. In the following, the procedures to calculate these parameters analytically are described. It may be noted that fracture mechanics deals with four important parameters. Out of these, two parameters may be termed global parameters: the load-line displacement (LLD) and crack mouth opening displacement (CMOD) of a specimen. Both parameters can be measured during an experiment using advanced instruments. The other two parameters may be termed crack tip local parameters: the crack tip opening displacement (CTOD) and J-integral. These local parameters are primarily applied to characterize a crack tip, that is, assess crack initiation and crack growth. These local parameters can be calculated using the global parameters CMOD or LLD. For this purpose, η-functions are used to correlate the local parameters with the global parameters.

The importance of η-functions is described in Chapter 2. A few important correlations are rewritten in this chapter for the ready reference of readers. To determine the η-function, the crack tip local parameters J and CTOD are expressed as the sum of the elastic and plastic components [1]:

$$J = J_{el} + J_{pl} \tag{6.1}$$

$$\delta = \delta_{el} + \delta_{pl}, \tag{6.2}$$

where J_{el} and J_{pl} represent the elastic and plastic components of the J-integral, respectively. Similarly, δ_{el} and δ_{pl} represent the elastic and plastic components

DOI: 10.1201/9781003310372-6

of the CTOD, respectively. The elastic components of the above equations are expressed as follows:

$$J_{el} = \frac{\left(1-\nu^2\right)K_I^2}{E} \tag{6.3}$$

$$\delta_{el} = \frac{\left(1-\nu^2\right)K_I^2}{mE\sigma_{ys}}, \tag{6.4}$$

where σ_{ys}, E, and ν are the yield stress, Young's modulus, and Poisson's ratio of a material, respectively. The parameters K_I and m are the stress intensity factor and an empirical plastic constraint factor, respectively. Part II of the present chapter provides a detailed procedure to determine the m-factor of a pre-cracked/notched small punch test (p-SPT) specimen.

The following four correlations [2] can be applied to calculate the plastic components of the J_{pl} and δ_{pl} as a function of the global parameters:

$$J_{pl} = \frac{\eta_{j,pl}^{LLD} A_{pl}^{LLD}}{b\,B_N} \tag{6.5}$$

$$J_{pl} = \frac{\eta_{j,pl}^{CMOD} A_{pl}^{CMOD}}{b\,B_N} \tag{6.6}$$

$$\delta_{pl} = \frac{\eta_{\delta,pl}^{LLD} A_{pl}^{LLD}}{b\,B_N \sigma_F} \tag{6.7}$$

$$\delta_{pl} = \frac{\eta_{\delta,pl}^{CMOD} A_{pl}^{CMOD}}{b\,B_N \sigma_F}, \tag{6.8}$$

where b, B_N, and σ_F are the remaining ligament $(W - a)$, net specimen thickness, and flow stress of the material, respectively. A_{pl}^{LLD} and A_{pl}^{CMOD} are the plastic components of the strain energy during the deformation of the specimen. These are determined by calculating the area under the curve between the load vs. LLD_{pl} plot or between load vs. $CMOD_{pl}$ plot, respectively. Hence, each fracture specimen can have four η-functions:

(i) $\eta_{J,pl}^{LLD}$, applied to correlate LLD_{pl} with J_{pl}
(ii) $\eta_{J,pl}^{CMOD}$, applied to correlate $CMOD_{pl}$ with J_{pl}
(iii) $\eta_{\delta,pl}^{LLD}$, applied to correlate LLD_{pl} with $CTOD_{pl}$
(iv) $\eta_{\delta,pl}^{CMOD}$, applied to correlate $CMOD_{pl}$ with $CTOD_{pl}$.

Readers may note the following two points:

(a) η-functions depend on the type of fracture specimen. This means that the η-functions are different for the fracture specimens' compact tension, TPB, and single-edge notched tension (SENT), for example. The correlations to calculate the η-functions for different types of fracture specimens are available in ASTM E1820 [3]. Readers are advised to go through the extensive literature available on this subject.
(b) There is no correlation available in the literature to calculate the η-function of a p-SPT specimen. The procedure to develop the correlation of the η-function of a p-SPT specimen is presented in Part I of this chapter. Readers can make use of such a procedure to derive the correlation of the η-functions of other fracture specimens of their interest.

In addition to the η-function, one parameter has a wide application in fracture mechanics. This parameter is popularly termed the m-factor. It correlates two local parameters: CTOD and J-integral. This correlation is expressed as follows:

$$J = m\,\sigma_F\ \delta, \tag{6.9}$$

where $\sigma_F = \left(\sigma_{ys} + \sigma_{UTS}\right)/2$ is the flow stress. The correlation to determine the m-factor once again depends on the type of fracture specimen. Such correlations of different fracture specimens are available in the literature and are summarized in Chapter 2. It may again be noted that such a correlation is not available for an p-SPT specimen in the literature. Part II of the present chapter is devoted to explaining a numerical procedure to derive the correlation of the m-factor of a p-SPT specimen. This procedure is generic in nature; hence, readers can make use of such a procedure to derive the correlation of the m-factors of other fracture specimens of their interest.

6.2 EFFECTIVE CRACK LENGTH AND LIGAMENT OF A p-SPT SPECIMEN AFTER MOUNTING ON EXPERIMENTAL FIXTURES

The complete physical length of the crack and ligament of any conventional fracture specimen remain pertinent during the fracture process of the specimen. However, in the case of a p-SPT specimen, the condition is somewhat different. The periphery of the p-SPT specimen is firmly mounted between two dies in the experimental set up to some width. Because of this, the length of the crack lying within this width of the specimen does not open up and hence does not participate in the fracture process of the specimen. Consequently, the effective length of the crack for the fracture mechanics analysis of the specimen is less than the total physical length of the crack. Fig. 6.1 shows the methodology to calculate the effective length of the crack and the effective ligament. These are shown as $a = X - 2.5$ mm and $W = 5$ mm

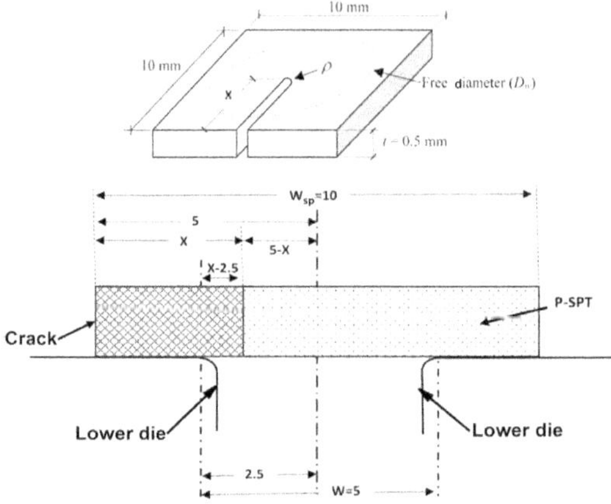

Unmounted p-SPT Specimen Physical Dimensions:

a. P-SPT specimen dimensions: 10 × 10 × 0.5 mm
b. W_{sp}: Dimension of specimen along crack length=10 mm
c. X: Length of the through-thickness crack

Effective p-SPT Specimen Dimensions when Mounted on Experimental Fixtures:

a. Effective length of the crack 'a'= X - 2.5 mm (This is the unsupported length of the crack lying within the cavity of the lower die having effective diameter W=5 mm).
b. Only unsupported length of the crack is expected to bend and open under the punch load.
c. The ligament of the unsupported crack is W.
d. The Following table shows the relation between X/W_{sp} and a/W for different physical dimensions of the crack.

Unmounted physical dimensions (ligament width W_{sp}=10 mm)		Effective dimensions when mounted on the experimental fixtures. (effective ligament width W=5 mm)	
Crack Length X (mm)	X/W_{sp}	Effective Crack Length a (mm)	a/W
4.00	0.400	1.50	0.30
4.25	0.425	1.75	0.35
4.50	0.450	2.00	0.40
4.75	0.475	2.25	0.45
5.00	0.500	2.50	0.50
5.50	0.550	3.00	0.60

FIG. 6.1 Procedure for calculating the effective crack length and effective remaining ligament of a p-SPT specimen when mounted on experimental fixtures

respectively in Fig. 6.1 for the specific dimensions of a p-SPT specimen. The table included in Fig. 6.1 also shows some physical dimensions and the corresponding effective dimensions of a p-SPT specimen considering the effect of mounting the specimen between the two dies. The entries in this table are used widely in Chapter 6 and Chapter 7 of the present book dealing with p-SPT specimens.

6.3 PART I: DEVELOPMENT OF THE $\eta_{plastic}$ CORRELATION TO EVALUATE THE CTOD OF THE p-SPT SPECIMEN

6.3.1 Analytical Procedure

A generic step-by-step procedure to derive the η_{pl}-function of the p-SPT specimen is as follows:

(a) Five stress-strain curves are first selected for the present analytical study. It may be noted that such sets of stress-strain curves cover the properties of most engineering materials of common interest. Fig. 6.2 shows these stress-strain plots.

(b) A p-SPT specimen with the dimensions 10 × 10 × 0.5 mm, as shown in Fig. 6.3, is considered to carry out the present numerical studies. Three sizes of crack lengths, that is, a/W = 0.3, 0.4, and 0.5, are considered. As explained in Section 6.2 of the present chapter, 'a' is the effective length of the crack, which is the unsupported length of the crack lying within the cavity of the lower die. The term 'W' represents the effective ligament and is equal to the diameter of the cavity of the lower die. The diameter of the cavity is shown in Fig. 6.3 using the notation 'D_w' and is equal to 5 mm. The load on the specimen is applied in displacement-controlled mode by incrementing the displacement of the punch in the vertical direction. Finite element (FE) analyses are carried out for 15 combinations representing five stress-strain data and three a/W ratios.

(c) Fifteen load vs. LLD data are thus numerically calculated using FE analyses. A sample plot of the calculated load vs. LLD is shown in Fig. 6.3. The nodal displacements of the FE analysis are subsequently post-processed to calculate the CTOD (δ) [4] as a function of the LLD. To calculate the CTOD, the definition given by Rice and presented in Chapter 7 is used. A sample plot is shown in Fig. 6.5.

(d) Subsequently, the load vs. LLD data are converted to load vs. plastic component of LLD denoted by LLD_{pl}. For this purpose, the elastic components of the LLD are subtracted from the total LLD. To obtain the elastic component of the LLD, it may be necessary to carry out elastic analyses of the p-SPT specimens for the a/W values of 0.3, 0.4, and 0.5 separately. A sample plot of load vs. LLD_{pl} up to the maximum loads is shown in Fig. 6.4.

(e) The area under load vs. LLD_{pl} is then calculated to obtain the plastic component of the strain energy A_{pl} as a function of the LLD data. This procedure is repeated for all 15 sets of FE analyses up to the maximum loads. A sample plot is shown in Fig. 6.6.

(f) The above procedure is also repeated to obtain CTOD (δ_{pl}) vs. LLD data for all 15 cases. A sample plot of δ_{pl} vs. LLD is shown in Fig. 6.7.

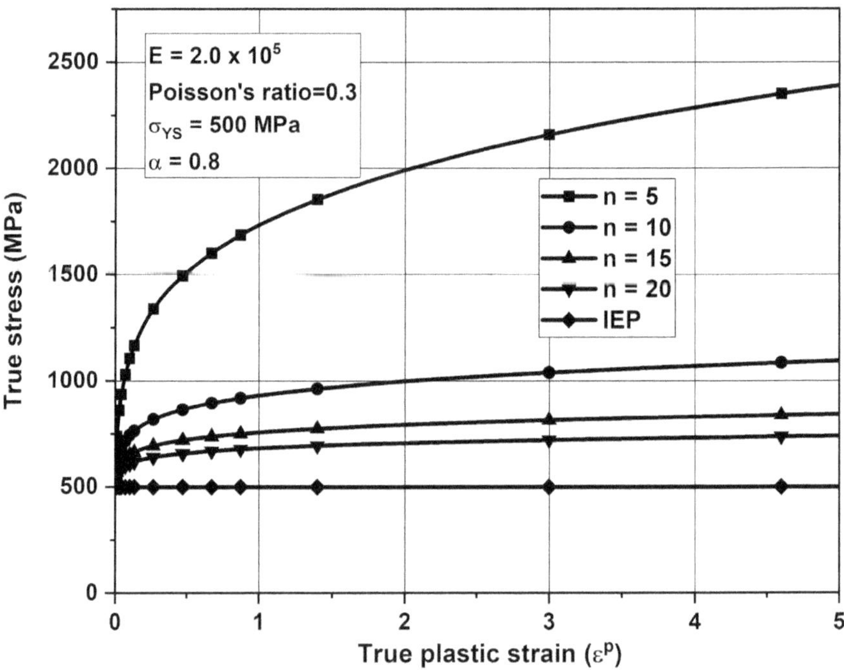

FIG. 6.2 Stress-strain data applied to carry out the analytical development of the η-function

(g) The two sets of data, that is, A_{pl} vs. LLD (Fig. 6.6) and δ_{pl} vs. LLD (Fig. 6.7) generated in step 'e' and step 'f' of each FE analysis are used to obtain a plot between $\frac{Apl}{B\sigma_F(W-a)^2}$ and $\frac{\delta pl}{(W-a)}$. A sample plot is shown in Fig. 6.8. It may be seen that, almost up to the maximum load, all the points in this plot lie close to a straight line. The slope of the best fit straight line of such a plot is the value of $\eta_{\delta,pl}^{LLD}$. This procedure is then repeated for all 15 FE analyses. This exercise leads to the 15 values of $\eta_{\delta,pl}^{LLD}$ for each combination of the material properties and a/W ratio.

(h) To develop a generalized correlation of $\eta_{\delta,pl}^{LLD}$ exclusively applicable to the p-SPT specimen, a set of curves between $\eta_{\delta,pl}^{LLD}$ and a/W for five material hardening coefficients are plotted. These plots are shown in Fig. 6.9. Readers may note the following points from these plots:

(i) The $\eta_{\delta,pl}^{LLD}$ function is almost independent of the material stress-strain data. This justifies the different forms of $\eta_{\delta,pl}^{LLD}$ reported earlier in the literature for other fracture specimens as being independent of the material parameters.

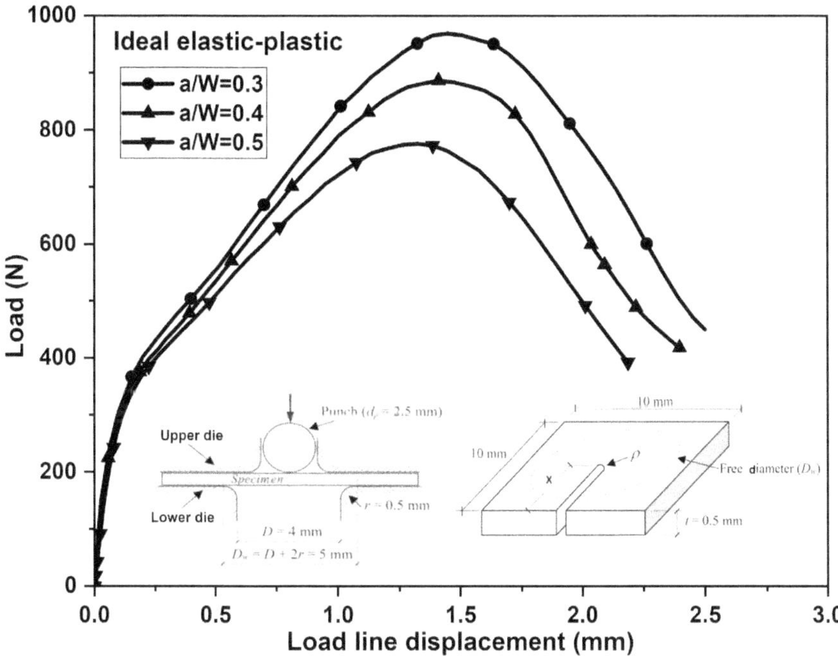

FIG. 6.3 Plot of load vs. LLD of p-SPT specimens. This figure also shows the salient dimensions of a p-SPT specimen used in the present analysis

(ii) The $\eta_{\delta,pl}^{LLD}$ of p-SPT is a strong function of crack parameter a/W. From the plots of Fig. 6.9, this correlation is derived as follows:

$$\eta_{\delta,pl}^{LLD} = 2.2778(a/W) - 0.4357. \tag{6.10}$$

(iii) Fig. 6.10 shows a comparison of the 15 computed $\eta_{plastic}$ parameters obtained using FE analyses and also calculated using the suggested correlation eq. (6.10). The newly developed empirical correlation is validated by the proximity of these points with the 45° straight line.

6.3.2 Verification of the Newly Developed $\eta_{\delta,pl}^{LLD}$ Correlation of a p-SPT Specimen against Experimental Data

An attempt is made in the section to verify the $\eta_{\delta,pl}^{LLD}$ correlation using the experimental data presented in Chapter 7. It is suggested that readers go through the experimental program reported in Chapter 7 before reading the remaining part of the present chapter to enable a better understanding of the content. The experimental load vs. LLD data of the p-SPT specimens of two structural steels are taken for this purpose. The experimental data are applied to calculate the values of A_{pl} using the procedure suggested above. In addition, eq. (6.10) is used to calculate η_{pl}. These two parameters

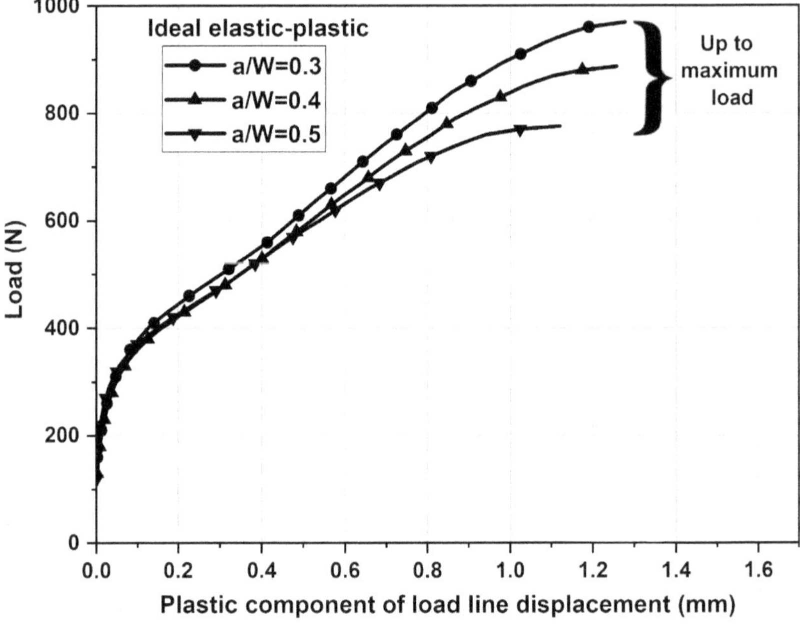

FIG. 6.4 Plot of load vs. LLD (plastic) up to the maximum load of p-SPT specimens

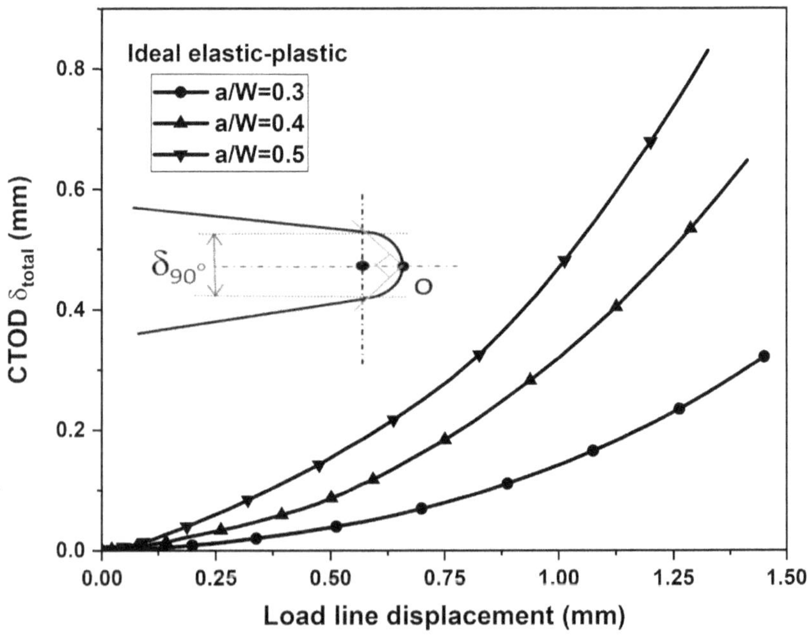

FIG. 6.5 Plot of CTOD (total) vs. LLD of p-SPT specimens

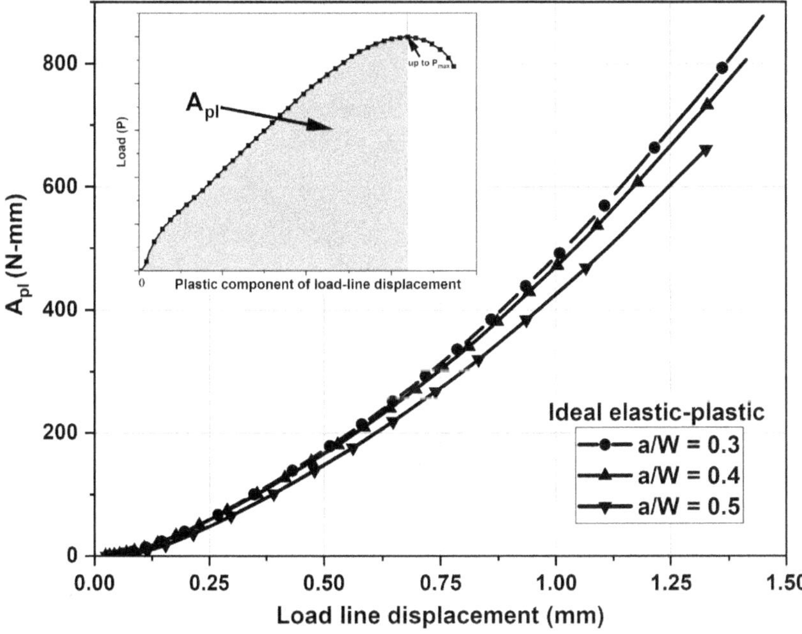

FIG. 6.6 Plot of plastic strain energy vs. LLD of p-SPT specimens. A procedure to calculate A_{pl} is also shown in this figure

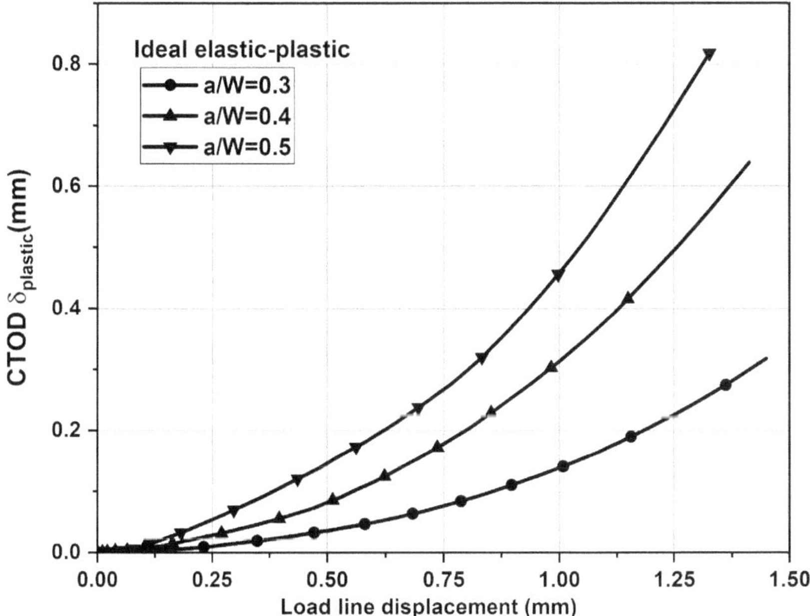

FIG. 6.7 Plot of CTOD (plastic) vs. LLD of p-SPT specimens

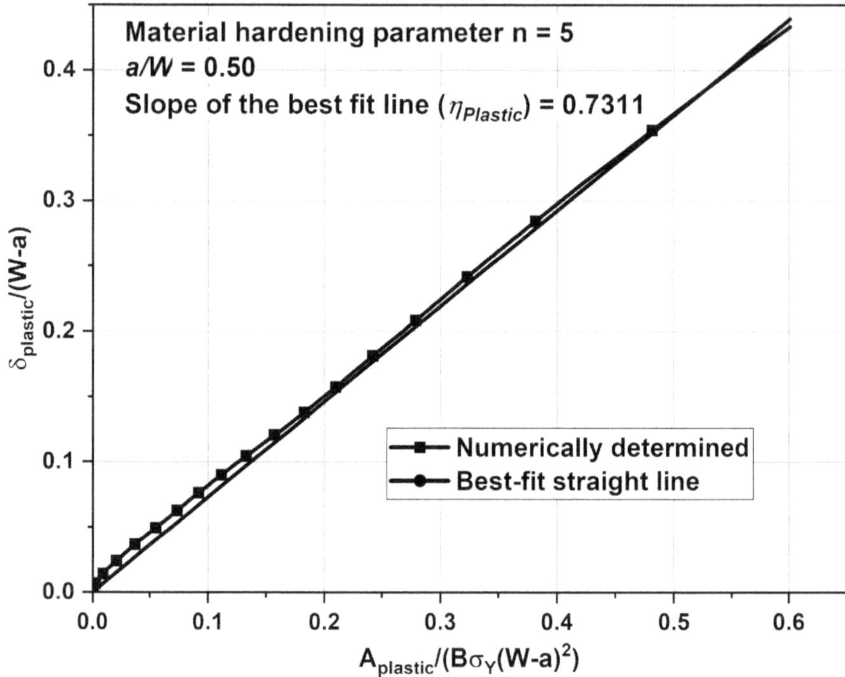

FIG. 6.8 Typical plot of $CTOD_{plastic}$ vs. plastic strain energy to calculate $\eta^{LLD}_{\delta,pl}$ by fitting a best fit straight line. Both the *X*-axis and *Y*-axis parameters are suitably non-dimensionalized

are then used to calculate $CTOD_{plastic}$ by applying eq. (6.7) to all the measured experimental data. For verification purposes, the CTOD as a function of the LLD is also calculated using FE analysis of the p-SPT specimen applying the material properties of both structural steels. Fig. 6.11 shows a comparison of the numerically calculated values of CTOD against the values calculated using the correlation shown in eq. (6.10) for T91 structural steels. A good comparison indicates the usefulness of the newly developed $\eta^{LLD}_{\delta,pl}$ correlation of the p-SPT specimen.

6.4 PART II: A NEW J-CTOD CORRELATION OF A p-SPT SPECIMEN CONVENTIONALLY DENOTED BY THE *m*-FACTOR

The derivation of a new *m*-factor exclusively applicable to the p-SPT specimen is described in the present section.

6.4.1 Development of a J-Integral and CTOD Database using Parametric Analysis of a p-SPT Specimen

An analytical procedure to develop a new empirical *m*-factor to evaluate the J-integral using the CTOD of a p-SPT specimen is presented here. It is expected

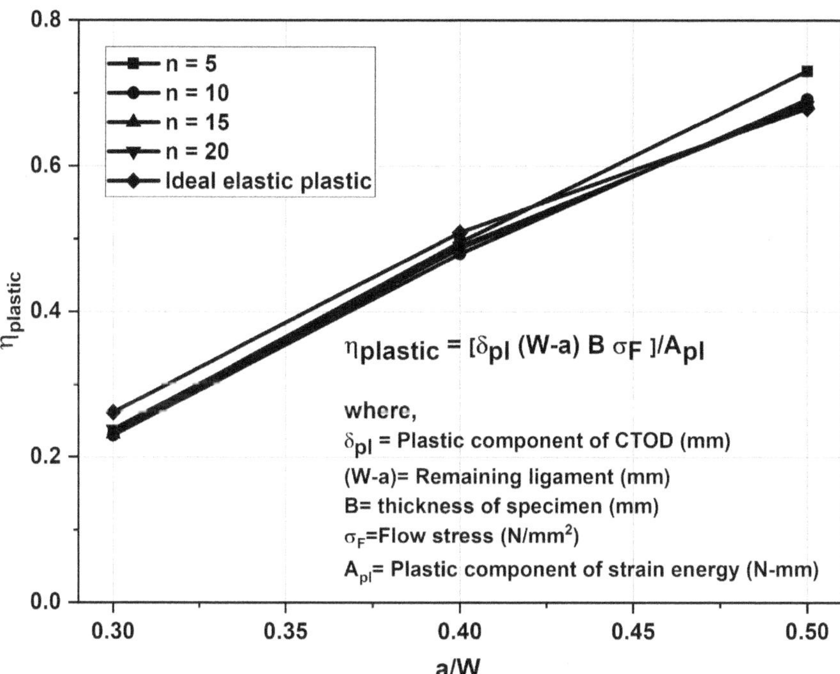

FIG. 6.9 Plot of $\eta_{plastic}$ vs. a/W for different values of hardening coefficients and a/w

that such a study would be useful for readers adopting an analogous analytical procedure to develop a new m-factor of any other specimen of their interest.

(a) Fifteen material stress-strain data are selected for the present analytical study. These data are obtained by different combinations of three yield stresses (300 MPa, 400 MPa, and 500 MPa) and five material strain hardening coefficients (5, 7, 10, 15, and 20). The stress-strain curves are generated for different combinations of yield stresses and hardening coefficients using the Ramberg–Osgood material model. A sample plot is shown in Fig. 6.12.

(b) As in Part I, the FE models of a square p-SPT specimen with dimensions of 10 × 10 × 0.5 mm and three values of a/W of 0.3, 0.4, and 0.5 are prepared, where 'a' is the effective crack length, which is the initial unsupported length of the crack lying within the cavity of the lower die and 'W' is the effective ligament, which is the diameter of the cavity of the lower die equal to 5 mm, as shown in Fig. 6.1. The load on the specimen is applied in displacement-controlled mode by incrementing the displacement of the punch in the vertical direction. Analyses are carried out for all combinations of 15 sets of material stress-strain data and three values of a/W ratios; hence, there are, in total, 45 FE analyses results.

(c) Using the results of the 45 FE analyses, the load vs. LLD data are obtained for further post-processing of the results. Three sample plots showing the

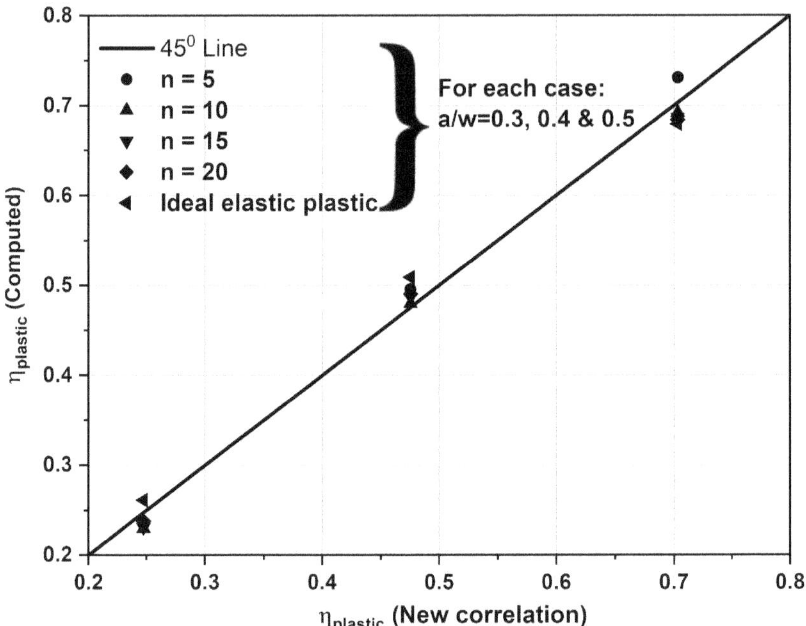

FIG. 6.10 Comparison of $\eta_{plastic}$ determined numerically and calculated using the newly developed correlation

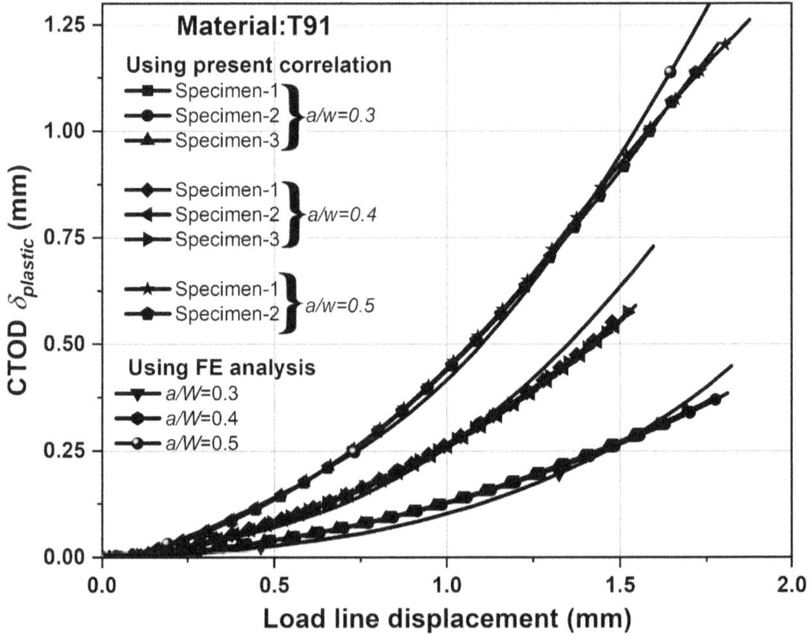

FIG. 6.11 Verification of newly developed correlation $\eta_{plastic}$ against experimental data

numerically calculated load vs. LLD are presented in Figs. 6.13(a), (b), and (c). It may be observed that the load vs. LLD data are strong functions of the yield stress, hardening coefficient, and a/W of the p-SPT specimen.

(d) The numerically calculated nodal displacements are also post-processed to calculate CTOD (δ) as a function of the LLD for each FE analysis. To calculate the CTOD, the definition presented by Rice and explained in Chapter 7 is used. To show the dependency of the CTOD on the input parameters, three sample plots are shown in Figs. 6.14(a), (b), and (c). The following conclusions may be drawn from these plots:

(i) The CTOD does not depend on the material stress-strain data, that is, the yield stress and hardening coefficient. This is illustrated in Figs. 6.14(a) and (b).

(ii) However, the value of the CTOD is a strong function of a/W, as shown in Fig. 6.14(c).

These two observations are important when deriving the expression of the *m*-factor of the p-SPT specimen.

(e) The computed stress-strain data from the FE analyses are then used to calculate the J-integral [5]. The domain integral method suggested in [6] is employed for this purpose. The J-integral is calculated for the first 15 contours surrounding

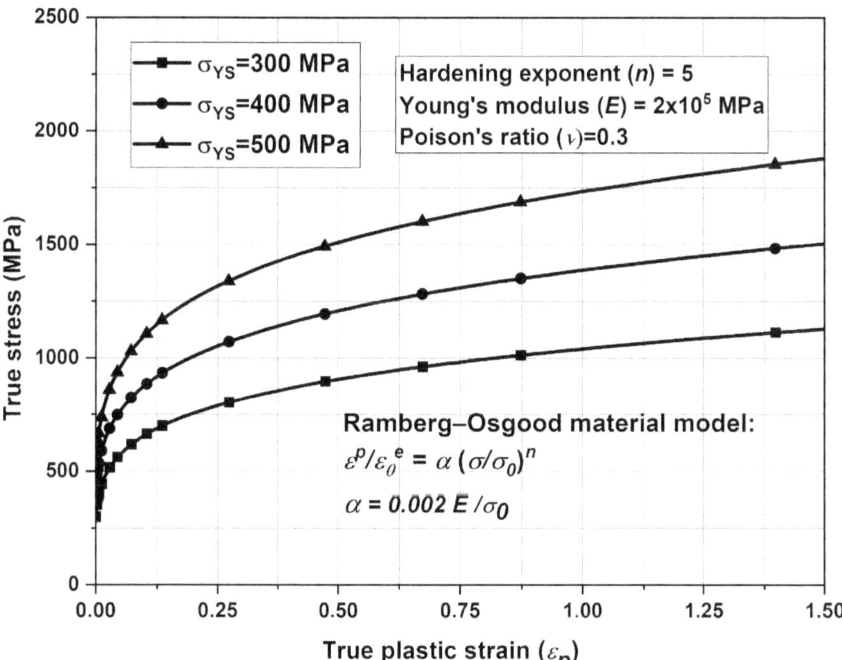

FIG. 6.12 Plot of true stress vs. true plastic strain used for developing the *m*-factor correlation of a p-SPT specimen

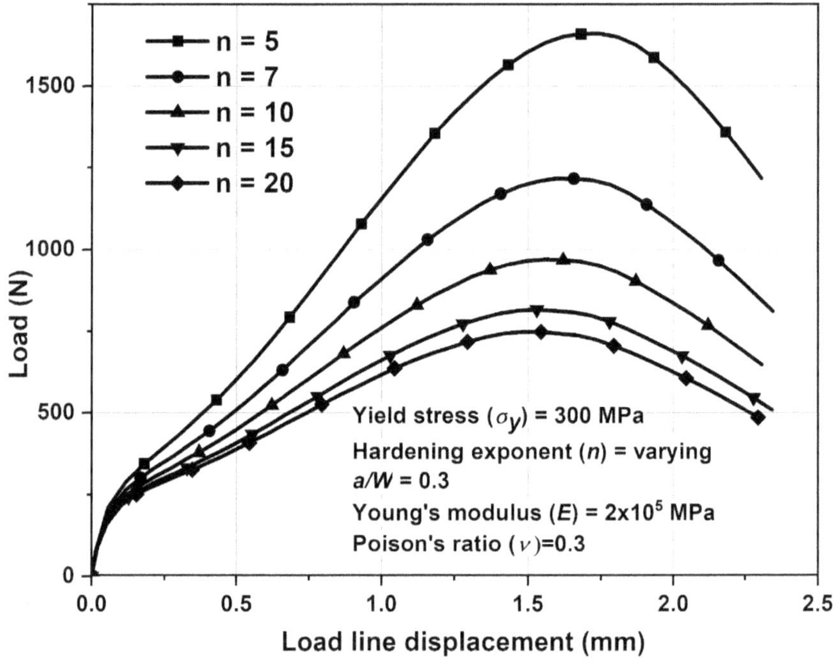

FIG. 6.13(a) Load vs. LLD of the p-SPT specimen for varying hardening coefficients

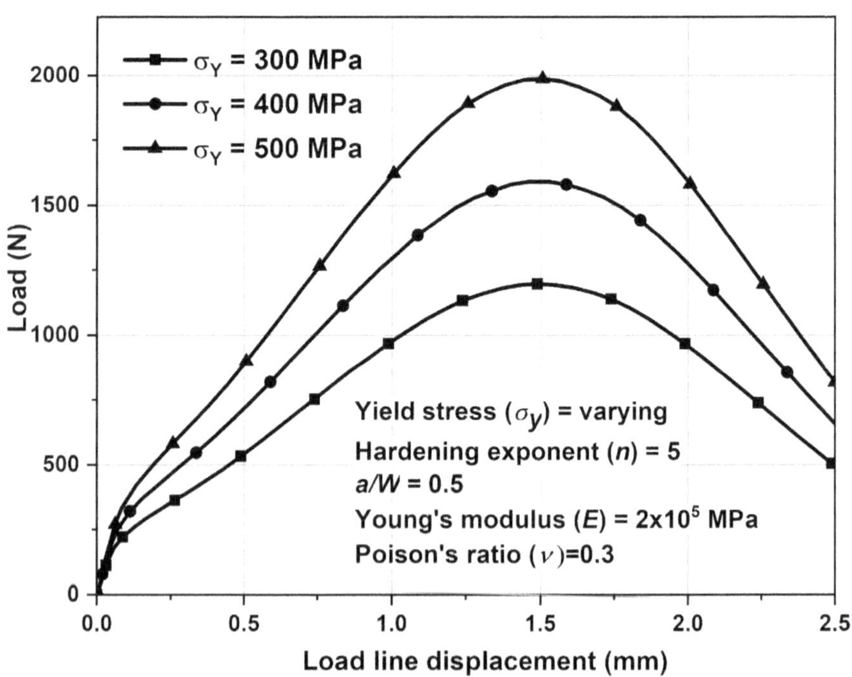

FIG. 6.13(b) Load vs. LLD of the p-SPT specimen for varying yield stresses

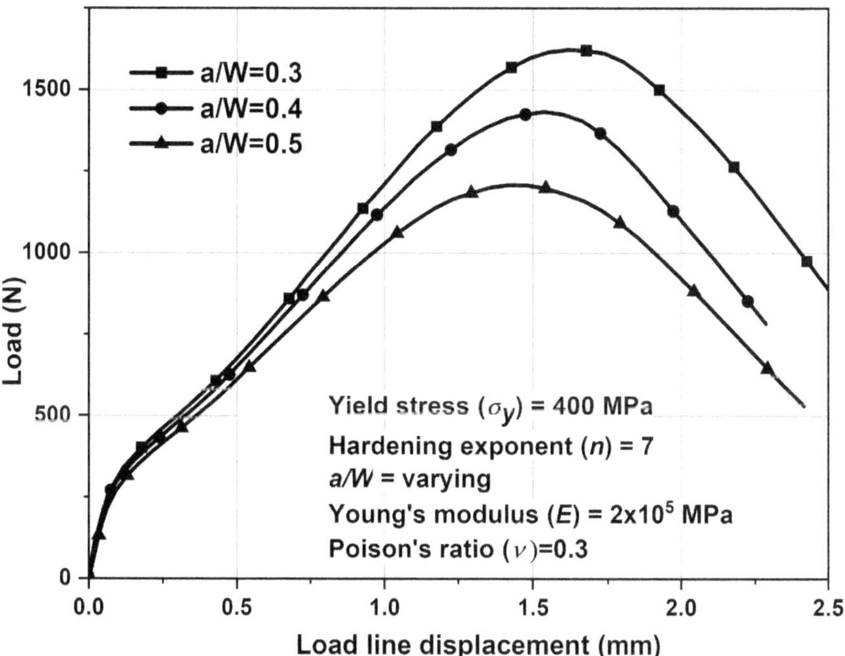

FIG. 6.13(c) Load vs. LLD of the p-SPT specimen for varying *a/W*

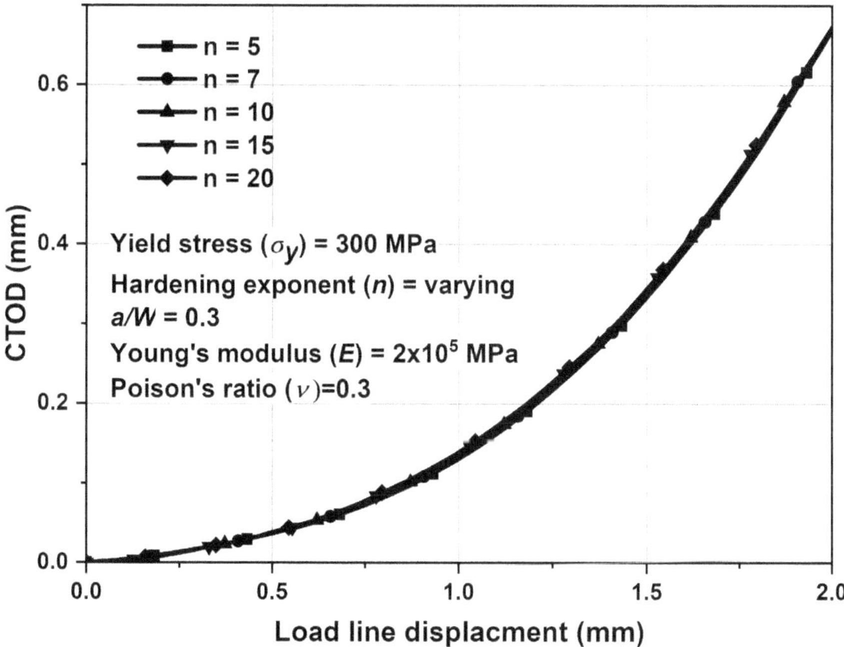

FIG. 6.14(a) CTOD vs. LLD of the p-SPT specimen for varying hardening coefficients

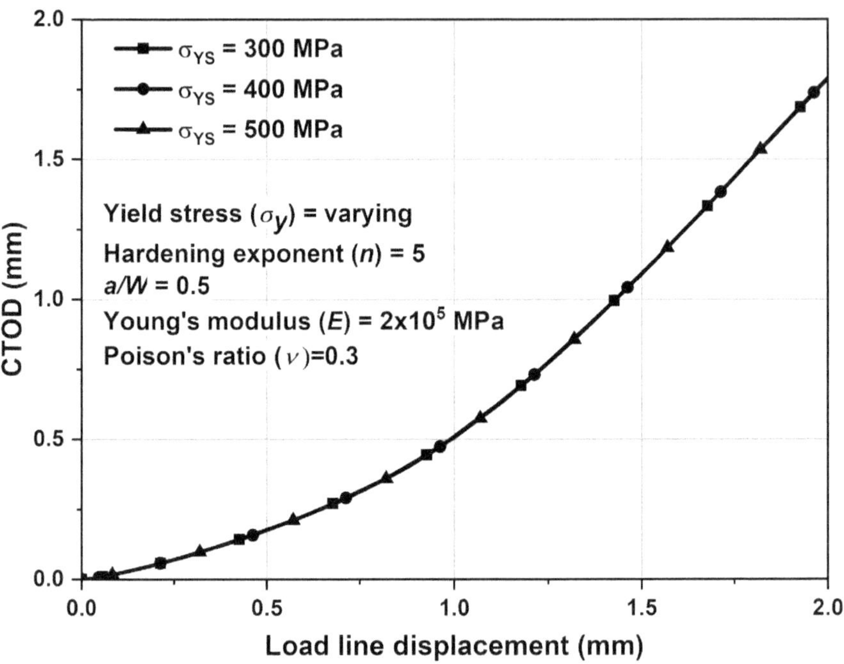

FIG. 6.14(b) CTOD vs. LLD of the p-SPT specimen for varying yield stresses

the crack tip of each analysis. It is observed that there is a minor loss of path independency at higher loads. Figs. 6.15 (a), (b), and (c) show the plots of the average J-integral as a function of the LLD for different strain hardening coefficients, yield stresses, and a/W , respectively. It may be observed from Fig. 6.15 (a–c) that, unlike the CTOD, the J-integral is a strong function of all three parameters, that is, the material yield stress, hardening coefficient, and a/W . On the other hand, readers may recall from point d above that the CTOD is a function of a/W only. Hence, to determine a function to correlate the J-integral with the CTOD is an involved task.

6.4.2 Development of the *m*-Factor to Correlate the J-integral and CTOD of a p-SPT Specimen

(a) The calculated J-integral is plotted against the CTOD of p-SPT specimens using the data shown in Figs. 6.14 (a–c) and 6.15 (a–c). Figs. 6.16 (a–c) show these plots. As expected, all the plots between the J-integral and CTOD are functions of σ_{ys}, n, and a/W . It may also be noted from Figs. 6.16 (a–c) that it may be necessary to have two best fit straight lines over the two ranges of the CTOD to correlate J with the CTOD for improved accuracy.

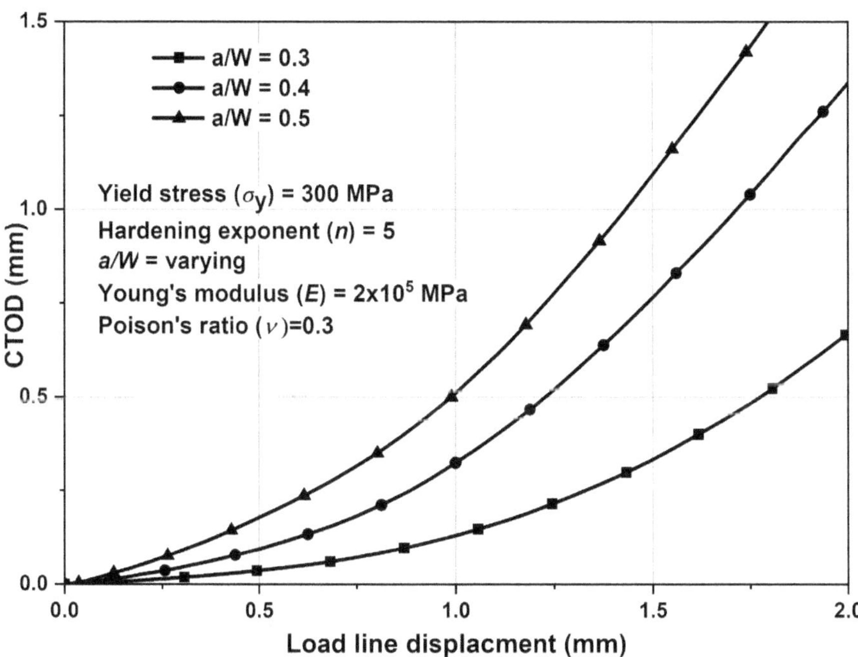

FIG. 6.14(c) CTOD vs. LLD of the p-SPT specimen for varying *a/W*

(b) The slopes of both these suggested best fit straight lines of all 45 FE analyses are calculated for each plot independently shown in Figs. 6.16 (a–c). These slopes are then converted to *m*-factors by dividing them by the flow stress σ_F of the corresponding material, as shown in eq. (6.9). Hence, in total, 90 *m*-factors are calculated using (a) the 15 sets of material properties by combining five yield stresses and three hardening coefficients, (b) the three sets of a/W, and (c) the two best fit straight lines for each J-CTOD plot.

(c) To derive empirical correlations of the *m*-factor of both best fit straight lines separately, the following equations are used:

- Over the range $0 \le \text{CTOD}\left(\delta\right) \le 0.10$ mm (with the condition that the first best fit straight line passes through CTOD = 0 of Fig. 6.16 plots),

$$m_1 = A_1 + B_1\left(\frac{\sigma_{ys}}{\sigma_{UTS}}\right) \tag{6.11a}$$

$$J = m_1 \sigma_F \delta.$$

- Over the range $0.10 < \text{CTOD} \le 1.0$ mm (with the condition that the second best fit straight line passes through CTOD = 0.1 mm of Fig. 6.16 plots),

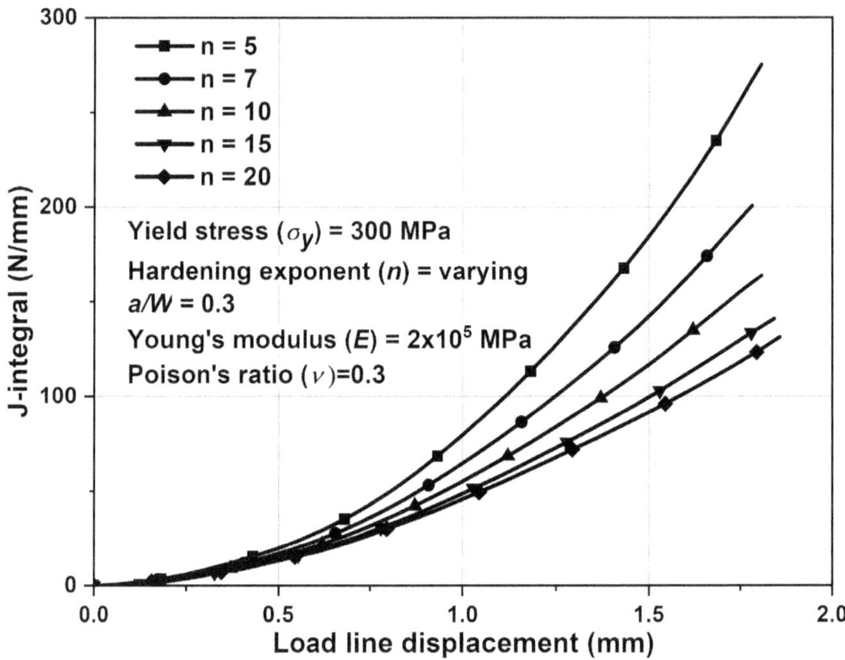

FIG. 6.15(a) Variation of the J-integral with the LLD for different hardening coefficients

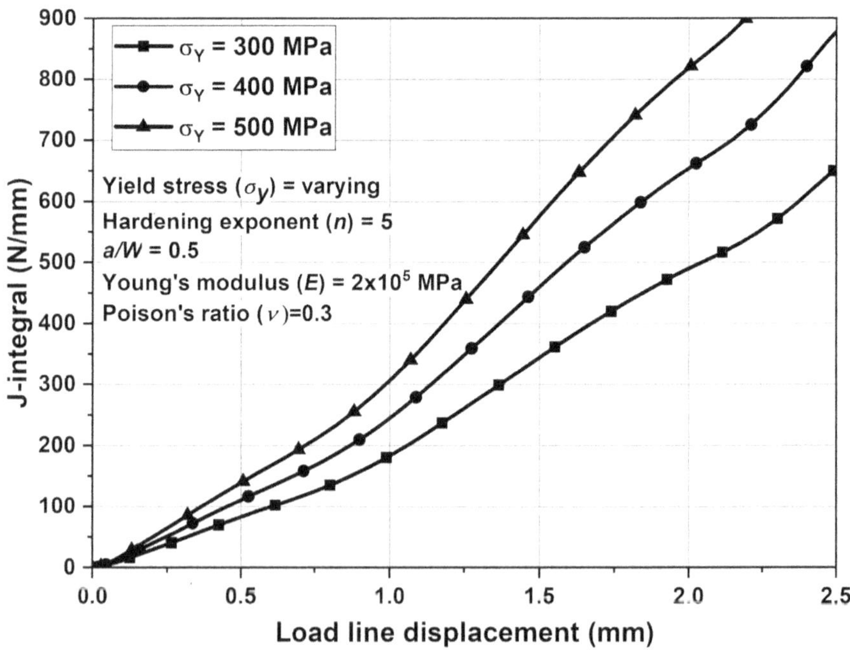

FIG. 6.15(b) Variation of the J-integral with the LLD for different yield stresses

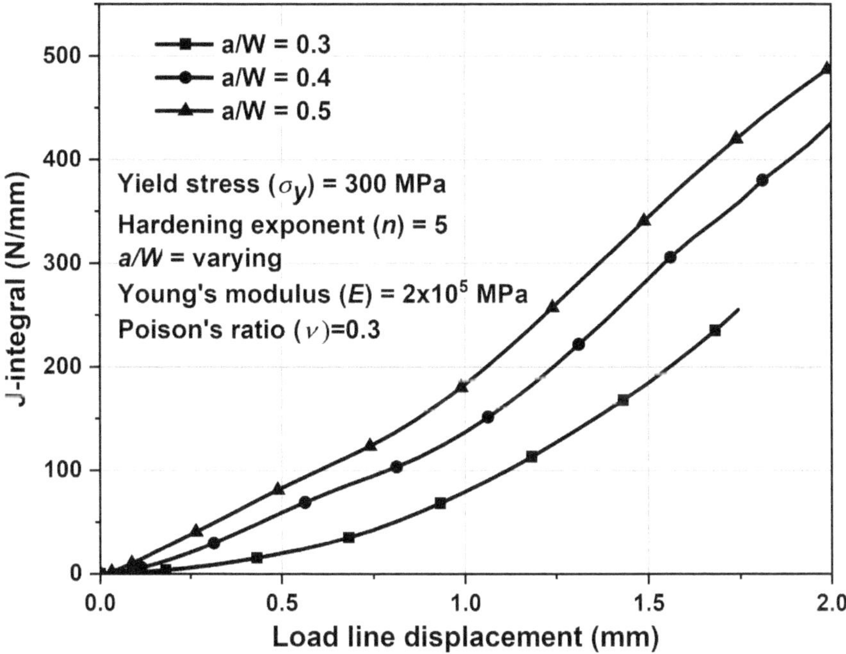

FIG. 6.15(C) Variation of the J-integral with the LLD for different a/W

$$m_2 = A_2 + B_2\left(\frac{\sigma_{ys}}{\sigma_{UTS}}\right) \tag{6.11b}$$

$$J = m_2\sigma_F\left(\delta - 0.10\right) + J_{0.10},$$

where $J_{0.10}$ is the value of the J-integral at CTOD = 0.10 mm derived from eq. (6.11a), that is, the intersection between the two best fit straight lines.

(d) It is observed that parameters A_i and B_i are functions of a/W. These parameters are derived using the values of the m-factor, that is, the slopes of both best fit straight lines. The variations of A_1 and B_1 as a function of a/W are shown in Fig. 6.17(a). Similarly, variations of A_2 and B_2 as a function of a/W are shown in Fig. 6.17(b). The corresponding empirical correlations are as follows;

$$A_1 = 2.9394 - 7.1315\left(\frac{a}{W}\right) + 8.495\left(\frac{a}{W}\right)^2$$

$$B_1 = -6.8881 + 34.37\left(\frac{a}{W}\right) - 43.945\left(\frac{a}{W}\right)^2 \tag{6.12}$$

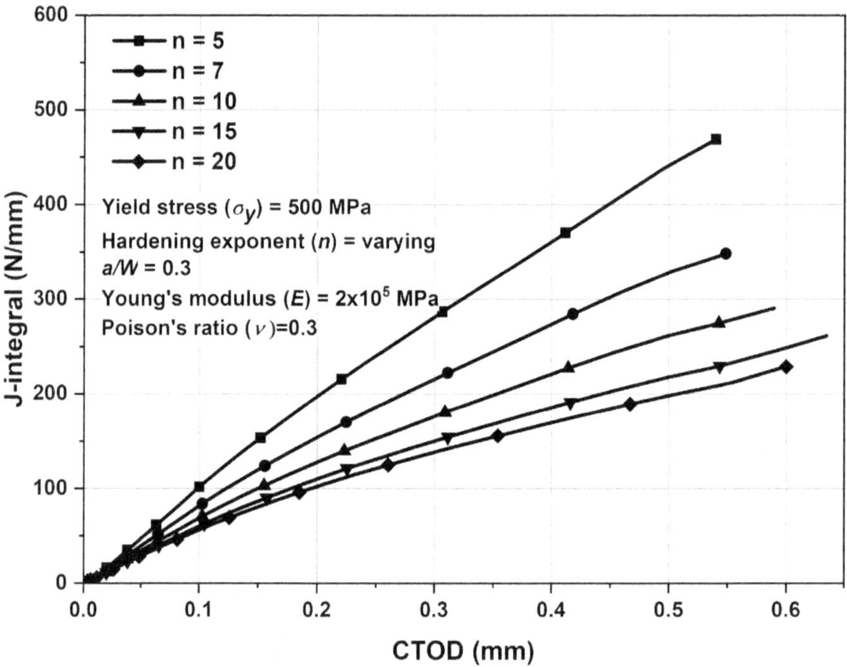

FIG. 6.16(a) Variation of the J-integral with the CTOD for different hardening coefficients

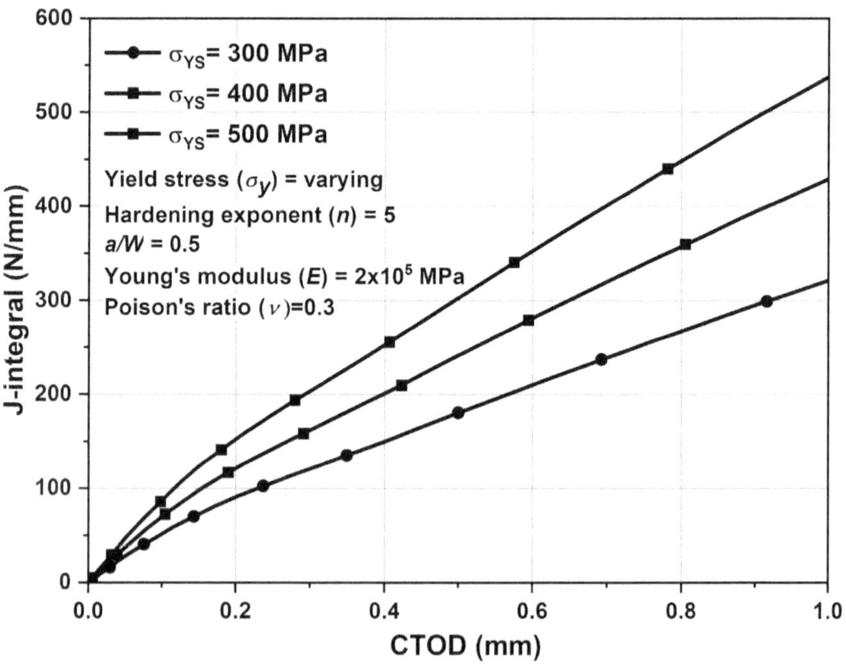

FIG. 6.16(b) Variation of the J-integral with the CTOD for different yield stresses

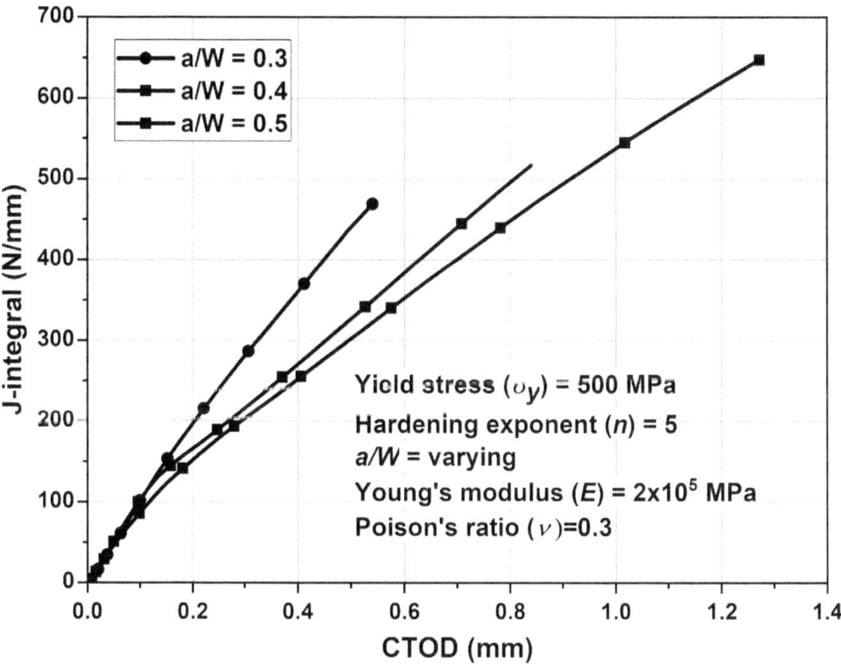

FIG. 6.16(c) Variation of the J-integral with the CTOD for different a/W

$$A_2 = 4.4397 - 12.679\left(\frac{a}{W}\right) + 11.425\left(\frac{a}{W}\right)^2 \tag{6.12}$$

$$B_2 = -0.3938 - 5.123\left(\frac{a}{W}\right) + 9.33\left(\frac{a}{W}\right)^2.$$

(e) Fig. 6.18 shows the plot of the *m*-factors calculated numerically against the values calculated using the newly proposed correlations, that is, eq. (6.11a), and eq. (6.11b). This figure also shows a 45° line. The accuracy of the newly developed empirical correlations of the *m*-factor of a p-SPT specimen are evident by the proximity of these points to the 45° straight line.

6.4.3 Application of Newly Developed Empirical Correlations of the *m*-Factor of p-SPT Specimens

In Chapter 7, the experiments conducted on p-SPT specimens of different a/W ratios are described. These specimens are fabricated using two structural steels: 20MnMoNi55 and T91. As the Ramberg–Osgood strain hardening exponent of these materials lies between 5 and 20, the newly developed *m*-factor correlations

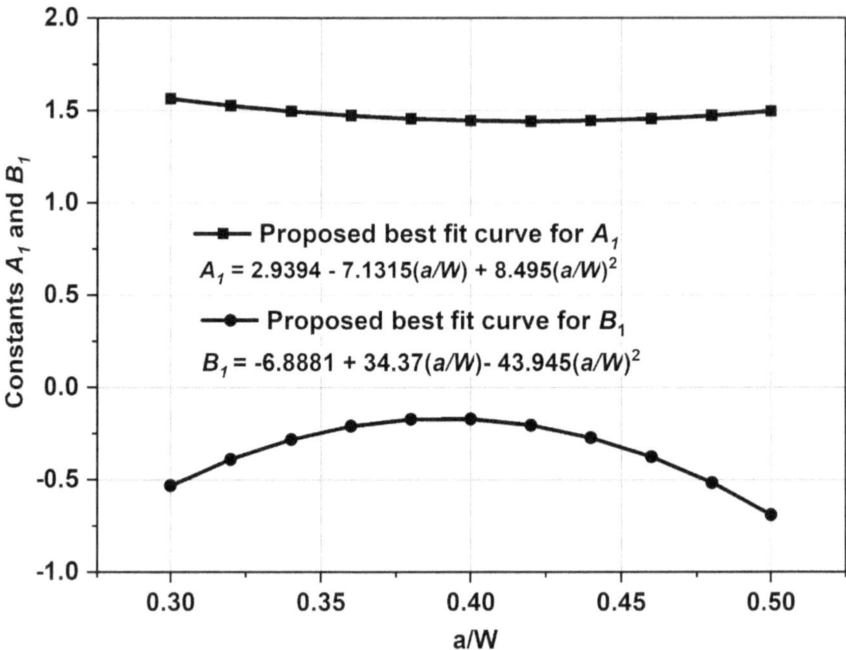

FIG. 6.17(a) Variations of empirical constants A_1 and B_1

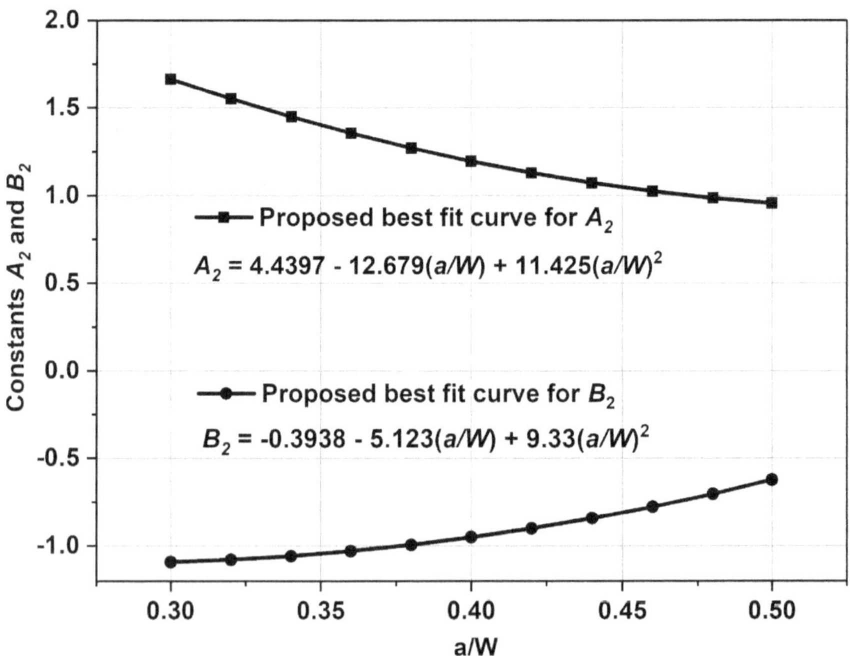

FIG. 6.17(b) Variations of empirical constants A_2 and B_2

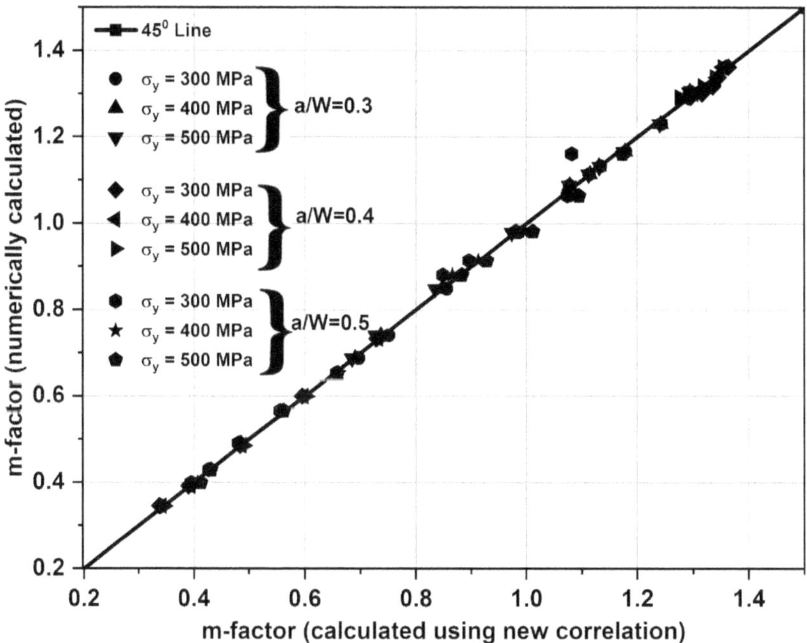

FIG. 6.18 Comparison of the *m*-factor calculated using a FE model and the newly proposed correlations

are applicable to both steels. The experimental load vs. LLD data are then used to calculate the $A_{plastic}$ of each load step. The newly developed $\eta_{\delta,pl}^{LLD}$ function is then applied to calculate the plastic component of $CTOD_{plastic}$ applying the values of $A_{plastic}$. The *m*-factor correlations shown in eq. (6.11) are then applied to calculate $J_{plastic}$ using the $CTOD_{plastic}$ of all tested specimens. Figs. 6.19 and 6.20 show the plots of $J_{plastic}$ as a function of the LLD of p-SPT specimens for different values of a/W of 20MnMoNi55 and T91 structural steel, respectively. For the sake of comparison, a set of FE analyses of these p-SPT specimens is also carried out, applying the stress-strain data of both structural steels. The FE outputs are then post-processed to calculate the J-integral at the crack tip of the p-SPT specimens. Fig. 6.19 and Fig. 6.20 also show the FE results. A good comparison of the FEM results with the results calculated employing the newly proposed correlation of the *m*-factor may be noted.

Readers may note that the advantage of the present correlation of the *m*-factor is that the *J*-integral can be calculated using experimental data directly without going through detailed FE analysis. Hence, the η-function and *m*-factor derived for the p-SPT specimens in the present chapter can be termed estimation techniques.

The correlations developed in the present chapter can also be used to calculate material fracture properties J_C corresponding to the crack tip stress triaxiality of the p-SPT specimen. For this purpose, the value of LLD should be noted at the instant of crack initiation during the experiment. This value of LLD_{ini} then can be applied to calculate $CTOD_{ini}$ by applying $\eta^{LLD}_{\delta,pl}$ derived in Part I of the present chapter, that is, eq. (6.10). $CTOD_{ini}$ can subsequently be used to calculate J_{ini} by applying the *m*-factor expression shown in eq. (6.11), derived in Part II of the present chapter.

6.5 CHAPTER CLOSURE

1. η-functions correlate two global parameters (i.e., LLD and CMOD) with the two local parameters (i.e., CTOD and J-integral) of a fracture specimen. Hence, there is the possibility of four sets of η-functions of a fracture specimen.
2. The *m*-factor correlates two crack tip local parameters (i.e., CTOD and J-integral).
3. The expressions of η-functions and *m*-factors depend on the type of fracture specimens.
4. The expressions of η-functions and *m*-factors for p-SPT specimens are not available in the literature. The purpose of the present chapter is to demonstrate numerical procedures to readers to derive such expressions for p-SPT specimens.

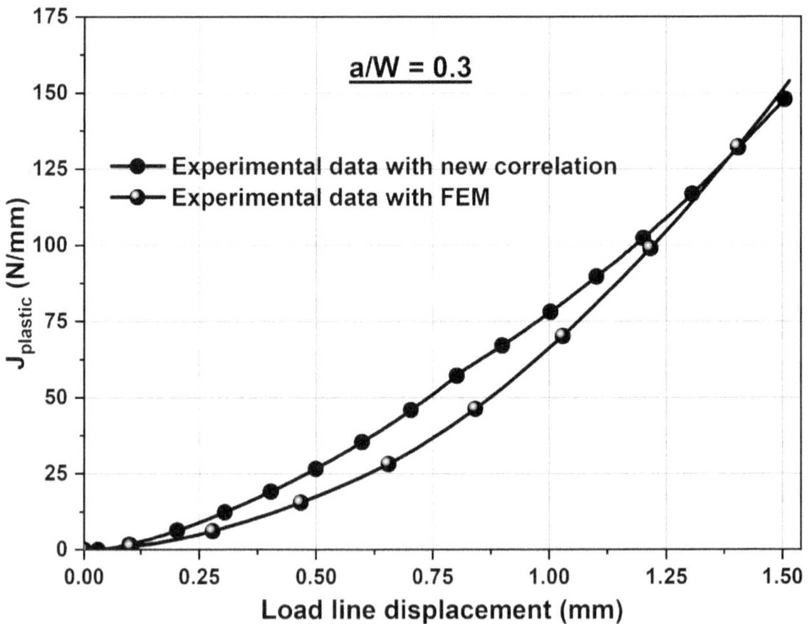

FIG. 6.19 Verification of the new *m*-factor correlation of the p-SPT specimen using the experimental data of 20MnMoNi55 structural material

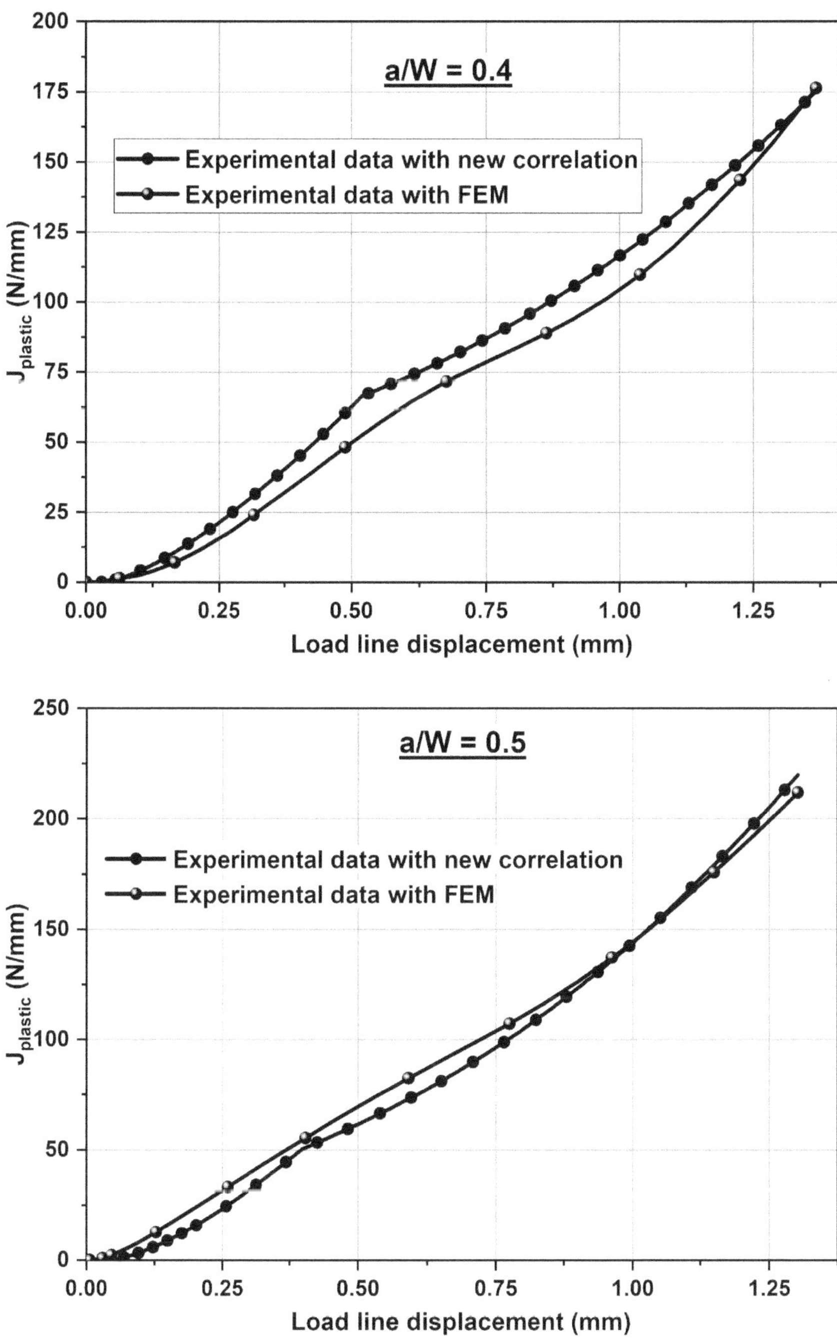

FIG. 6.19 (Continued)

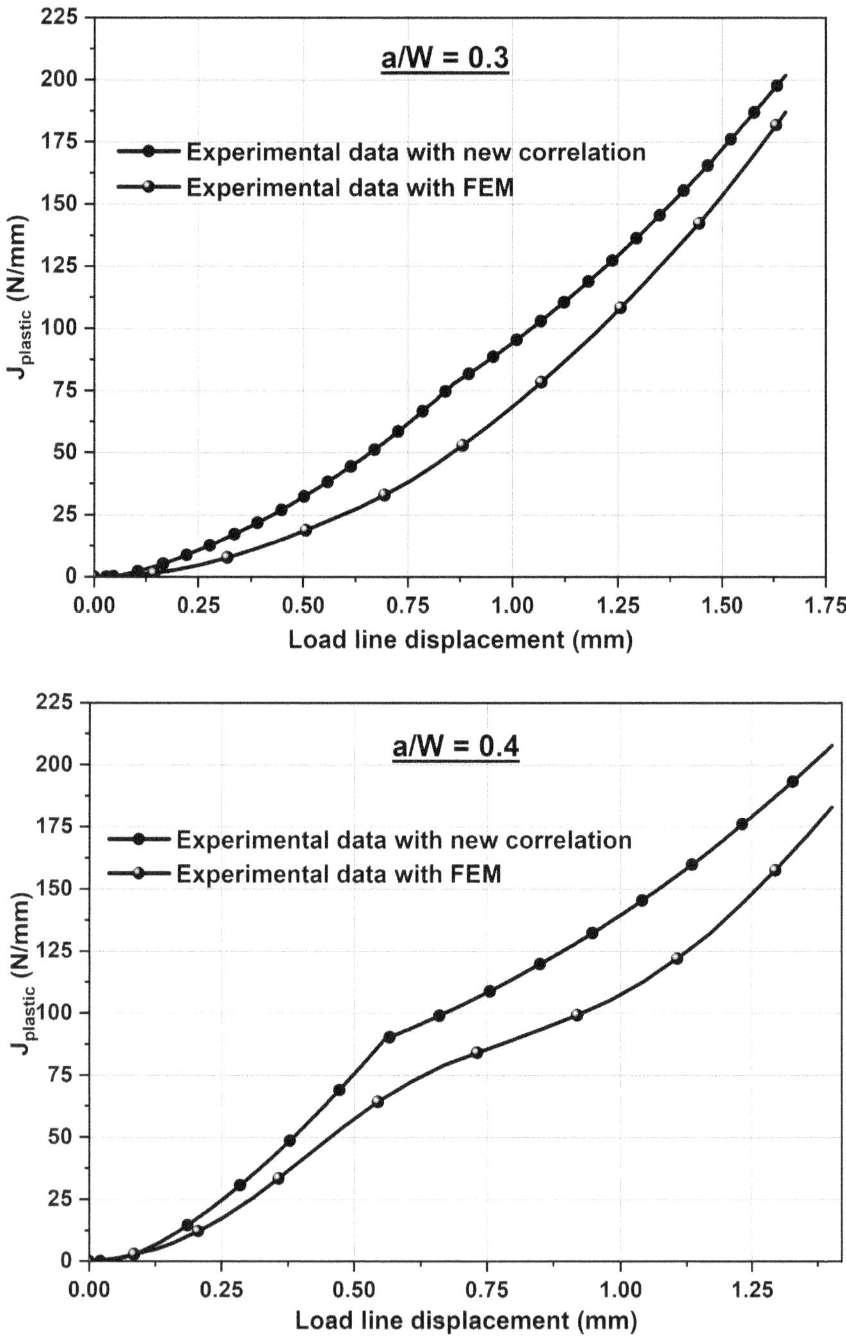

FIG. 6.20 Verification of the new *m*-factor correlation of the p-SPT specimen using the experimental data of T91 structural material

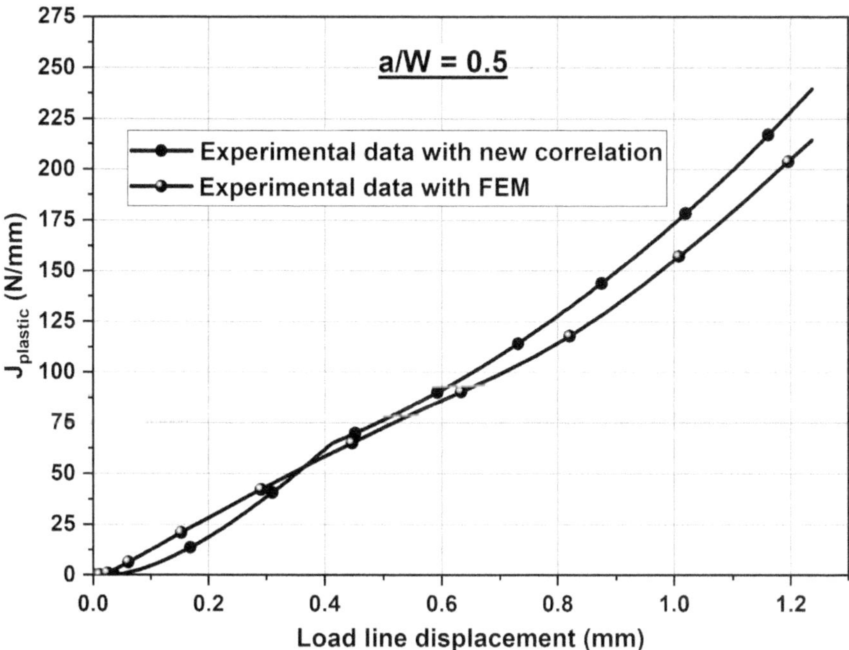

FIG. 6.20 (Continued)

5. A series of stress-strain curves are selected by varying the yield stress and hardening coefficient. The Ramberg–Osgood material model is then applied to calculate the complete stress-strain data.
6. The FE models of a p-SPT specimen with three a/W parameters are analyzed using the stress-strain data developed above. The expressions of the η-function and m-factor are then derived by post-processing the FE results.
7. It is observed that a single expression of the η-function of the p-SPT specimen is adequate to calculate the CTOD at the crack tip using experimental LLD data.
8. On the other hand, it is observed that a set of two linear expressions are essential for the m-factor to calculate the J-integral at the crack tip as a function of the CTOD. These two linear expressions are valid over the two ranges of the CTOD and are functions of the a/W ratio of the p-SPT specimen.
9. The derived correlations of η-functions and m-factor are then validated against experimental data for the p-SPT specimens.
10. It is expected that the present chapter will be helpful to readers developing similar new correlations of any other fracture specimen of their choice for which η_{pl} and m-factor correlations are not available in the literature.

REFERENCES

[1] Kirk MT, Dodds RH. J and CTOD estimation equations for shallow cracks in single edge notch bend specimens. Journal of Testing and Evaluation, 21 (1993). 228–38.
[2] Wang E, De Waele W, Hertelé S. A complementary η_{pl} approach in J and CTOD estimations for clamped SENT specimens, Engineering Fracture Mechanics,147 (2015), 36–54.
[3] ASTM Standard E1820-11. Standard Test Method for Measurement of Fracture Toughness, Book of Standards Volume: 03.01 (2009), DOI: 10.1520/E1820-11.
[4] Kirk MT, Wang Y-Y. Wide range CTOD estimation formulae for SE (B) specimens, ASTM STP16382S (1995) 126-141, DOI: 10.1520/STP16382S
[5] Rice JR. A path independent integral and the approximate analysis of strain concentration by notches and cracks. J Appl. Mech. 1968, 35, 379–86.
[6] Shih CF, Moran B, Nakamura T. Energy release rate along a three-dimensional crack front in a thermally stressed body. International Journal of Fracture, 30 (1986) 79–102.

7 A New Methodology to Calculate the J-R Curve of Structural Steels using p-SPT Data

Iaslim D. Shikalgar, B.K. Dutta, and J. Chattopadhyay

7.1 OVERVIEW

As mentioned in the previous chapters, the primary purpose of using small punch test (SPT) specimens is to determine the tensile and fracture properties of industrial materials subjected to in-service aging in industries dealing with hazardous processes, such as chemical plants and nuclear plants. Such specimens are tested largely to assess changes in material properties during service leading to an evaluation of changes in the safety margins of plant components over the years. SPTs are performed on miniaturized circular or square specimens with a 0.25–0.50 mm thickness. Such specimens of aged materials can be obtained by scrapping aged/irradiated industrial plant components without affecting the integrity of the components. Such a small volume of hazardous materials also can be handled without compromising the safety of plant personnel during material testing. Such a material testing methodology has found large-scale applications in many other areas where conventional tensile and fracture tests as per ASTM standards have practical difficulties. Such tests have numerous advantages, for example: (a) the tests can be used to monitor changes in the material properties of industrial plant component materials that are in service, (b) the experiments are simple to perform, (c) the tests can be used to determine the material properties in cases in which voluminous materials are harmful to handle, (d) the tests can be used for the speedy screening of the procured materials to check the variation in the properties of bulk materials using random sampling, (e) the tests can be used to study weld joints and coatings, (f) the tests can be used to assess the residual life of industrial plant components while carrying out a life extension program, (g) the tests are very useful during the development of new alloys and exotic materials, and (h) the tests can be used for the failure analysis of engineering plant components. SPTs have recently gained popularity as an acceptable test method for obtaining elastic-plastic

DOI: 10.1201/9781003310372-7

material properties. The CEN published a European Code of Practice on Small Punch Tensile and Fracture tests in 2006, which was revised in 2007 [1].

As explained in previous chapters, these experiments are performed by clamping an SPT specimen between two dies and then punching the specimen centrally with a rigid spherical ball until a fracture occurs. During the experiment, the measured load vs. displacement data are recorded. The measured test data are then used to calculate the initial stiffness, yield load, maximum load, maximum displacement at failure, and area under the load vs. displacement curve. These data can then be used to estimate various material properties, for example, the yield stress, ultimate stress, biaxial fracture strain, and fracture toughness. The estimation of the material properties using the measured data is performed by applying empirical correlations developed by various researchers over many years. In Chapters 3, 4, and 5 of the present book, some such advanced correlations are derived, and case studies are presented to show their usefulness. The methodologies presented in these chapters can be used by readers to derive similar correlations for other materials and specimens.

Several efforts have been made over the last few decades to quantify the fracture properties of materials using conventional SPT specimens by developing various empirical equations. Such empirical equations correlate material fracture toughness with the biaxial fracture strain. These correlations are material-dependent and are developed mostly by deriving best fit equations on large sets of experimental data for a specific type of material [2]. Readers may refer to some of these correlations presented in Chapter 2. However, without an initial crack in the specimen, the evaluation of material fracture properties using such empirical correlations lacks some degree of confidence. Recently, several authors [3–8] introduced the initial crack (or notch) in SPT specimens. These SPT specimens are called pre-cracked SPT (p-SPT) specimens. Some of the empirical correlations developed for p-SPT specimens are also described in Chapter 2.

The present chapter will acquaint readers with a novel hybrid methodology to determine the fracture properties of structural materials using experimental p-SPT data. This hybrid methodology involves two steps: (a) the generation of experimental data of p-SPT specimens and (b) the finite element (FE) elastic-plastic analysis of the experimental data to obtain the J-R curve of the material. The development, implementation, and verification of this new methodology will be demonstrated in subsequent sections as Part I with the help of two real-life case studies. The case studies will be presented in four sections: (i) compositions and properties of the materials used in the case studies, (ii) fabrication and testing of the p-SPT specimens, (iii) elastic-plastic FE analysis of the p-SPT experimental data, and (iv) study of the experimental data in association of the elastic-plastic results to derive the J-R data of the material. Part II of the present chapter deals with the damage mechanics analysis of the p-SPT specimen using the Gurson–Tvergaard–Needleman (GTN) model to generate the material J-R curve. The purpose of this part is to validate the newly developed hybrid methodology using such an advanced computational method.

7.2 PART I: DEVELOPMENT AND APPLICATION OF A NEW HYBRID METHODOLOGY TO CALCULATE THE MATERIAL J-R CURVE USING p-SPT EXPERIMENTAL DATA

7.2.1 Materials

In this section, the development and application of the new hybrid methodology will be demonstrated through two case studies. Two widely used structural steels are considered for this purpose: 20MnMoNi55 and T91 structural steels. These materials have excellent mechanical and fracture properties. These materials have wide applications as structural materials in some front-line industries, such as nuclear industries. Table 7.1 shows the chemical compositions of these two structural steels.

The elastic-plastic properties of both materials may be expressed by the Holloman power law representing true stress (σ) vs. true plastic strain (ε_p) data, as follows:

$$\sigma = K\varepsilon_p^n, \tag{7.1}$$

where K is the strength coefficient and n is the strain hardening exponent. Table 7.2 shows the elastic-plastic properties of both materials at ambient temperature. Readers may note that, while the stress-strain curve is represented by the Ramberg–Osgood relation, the value of the hardening coefficient has the value of $1/n$ of both materials.

TABLE 7.1
Chemical compositions (in weight percentage) of 20MnMoNi55 and T91 structural steels

Material	C	Mn	Mo	Ni	Si	Cr	Al	N	P	S	Fe
20MnMoNi55	0.19	1.33	0.53	0.52	0.23	0.062	0.056	0.009	0.007	0.001	Bal
T91	0.099	0.43	0.96	0.24	0.32	8.80	-	-	0.02	-	Bal

TABLE 7.2
Material properties of both structural steels at ambient temperature

Material	E (MPa)	σ_{YS} (MPa)	σ_{UTS} (MPa)	% Elongation	K (MPa)	n
20MnMoNi55	203000	420	586	22.6	927.10	0.1521
T91	203000	544	684	22.0	921.55	0.0848

7.2.2 Fabrication of p-SPT Specimens

This section briefly describes the fabrication methodology of pre-cracked SPT specimens. A typical square pre-cracked SPT specimen has dimensions of the order of 10 × 10 × 0.5 mm, as shown in Fig. 6.1. It is possible to fabricate specimens with varying initial crack lengths. The guidelines to fabricate p-SPT test fixtures and the procedure to test p-SPT specimens are available in the CEN code [1]. Readers are advised to read this document. The fabrication of p-SPT specimens is a skilled job. High-precision micromachining and laser-induced micromachining techniques need to be used [7]. A through-thickness uniform crack and a minimum crack tip radius of the order of 0.1 mm are achievable using these fabrication techniques. As part of the present case studies, p-SPT specimens are fabricated from a block of material provided by the nuclear authorities of India. The wire electrical discharge machining technique is first used to slice a bar from the block of material to ensure a uniform shape and depth. Subsequently, the desired length of the crack is obtained in the bar. The bar is then sliced to generate p-SPT specimens. Mechanical polishing of both surfaces of the specimens is carried out using P600 to P2400 grit size abrasive paper. Three sets of fabricated specimens have through-thickness initial crack lengths of 4.0, 4.5, and 5.0 mm. As explained in Chapter 6, the effective crack length in the case of the p-SPT specimen is different from the physical crack length. As shown in Fig. 6.1, the effective crack lengths in the present cases are 1.5, 2.0, and 2.5 mm and the corresponding a/w are 0.3, 0.4, and 0.5 respectively.

The fabricated specimens are mounted between lower and upper dies, and loaded centrally by a punch with a rigid hemispherical ball for testing. The test setup is fabricated as per the recommendations of the CEN code [1]. The SPTs are performed at room temperature using a high-precision machine.

7.2.3 Experimental Program for p-SPT Specimens

All six sets of p-SPT specimens are tested. Every set has three specimens of equal crack lengths and the same material. Hence, in total, 18 specimens are tested. All tests are carried out at ambient temperature. A digitally controlled 5.0 kN capacity servo-hydraulic testing machine is used for this purpose. The force acting on the specimen is measured using the load cell and the punch displacement is measured using the linear variable differential transducer. The tests are conducted under displacement-controlled loading with a constant displacement velocity of 0.2 mm/min. The experimental loads vs. punch displacement plots of both structural steels are shown in Figs. 7.1 (a–c) and Figs. 7.2 (a–c), respectively. Readers may note that the shape of every plot clearly indicates the elastic deformation, elastic-plastic transition zone, generalized plastic zone, plastic instability, plastic softening, and final failure regions. The experimental data generated in this section are subsequently used to obtain material fracture properties adopting the new methodology described below. The purpose of the present chapter is to show readers this new methodology and its verification.

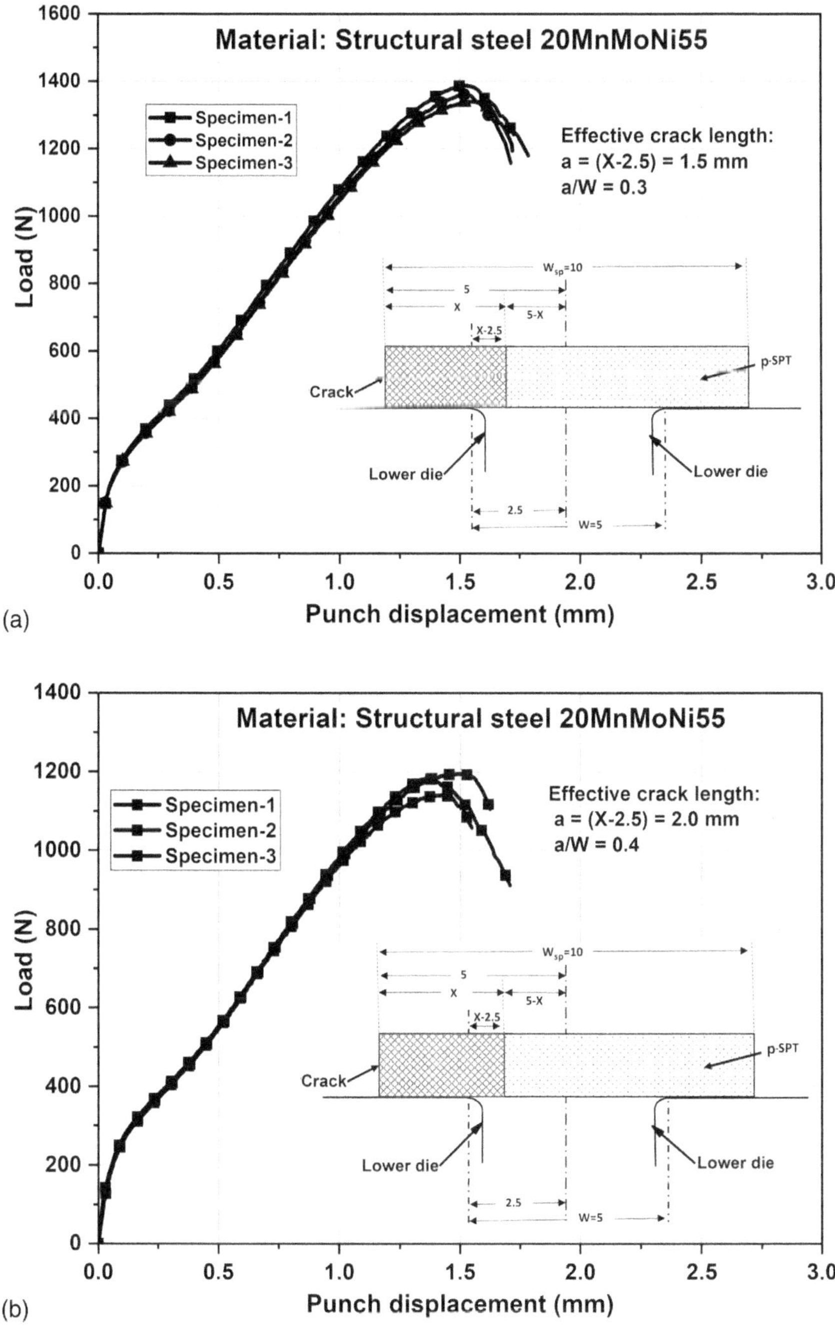

FIG. 7.1 Experimental load vs. load-line-displacement plots of p-SPT specimens of 20MnMoNi55 structural steel for *a/W* = (a) 0.30, (b) 0.4, and (c) 0.50

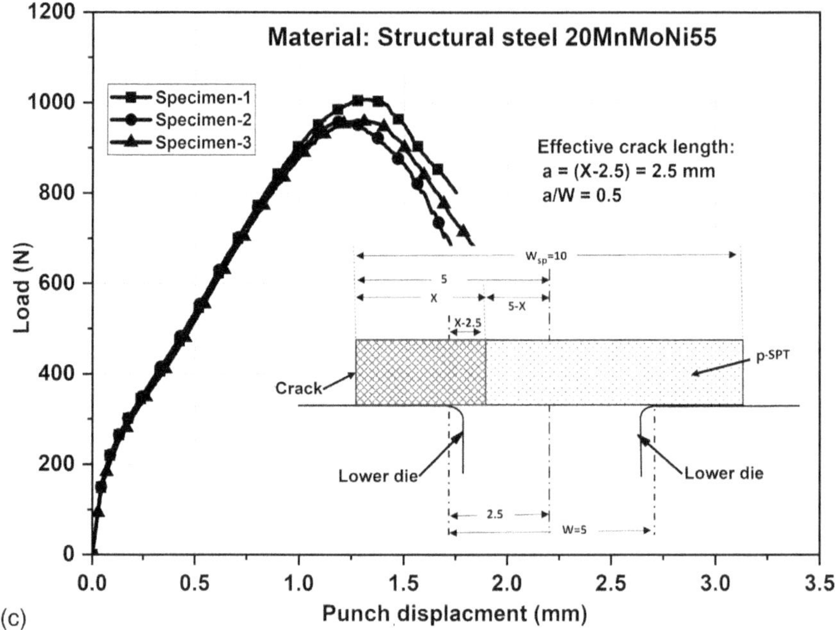

FIG. 7.1 (Continued)

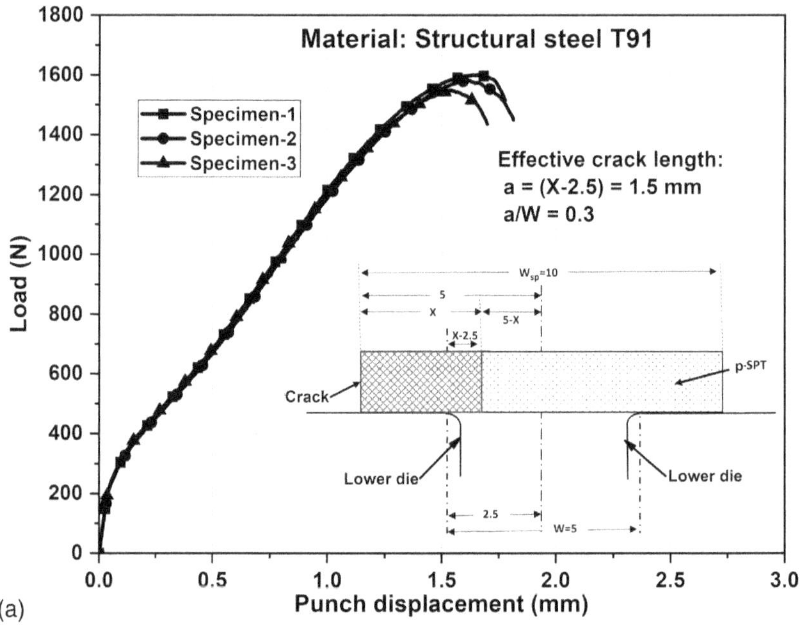

FIG. 7.2 Experimental load vs. load-line-displacement plots of p-SPT specimens of T91 structural steel for a/W = (a) 0.30, (b) 0.4, and (c) 0.50

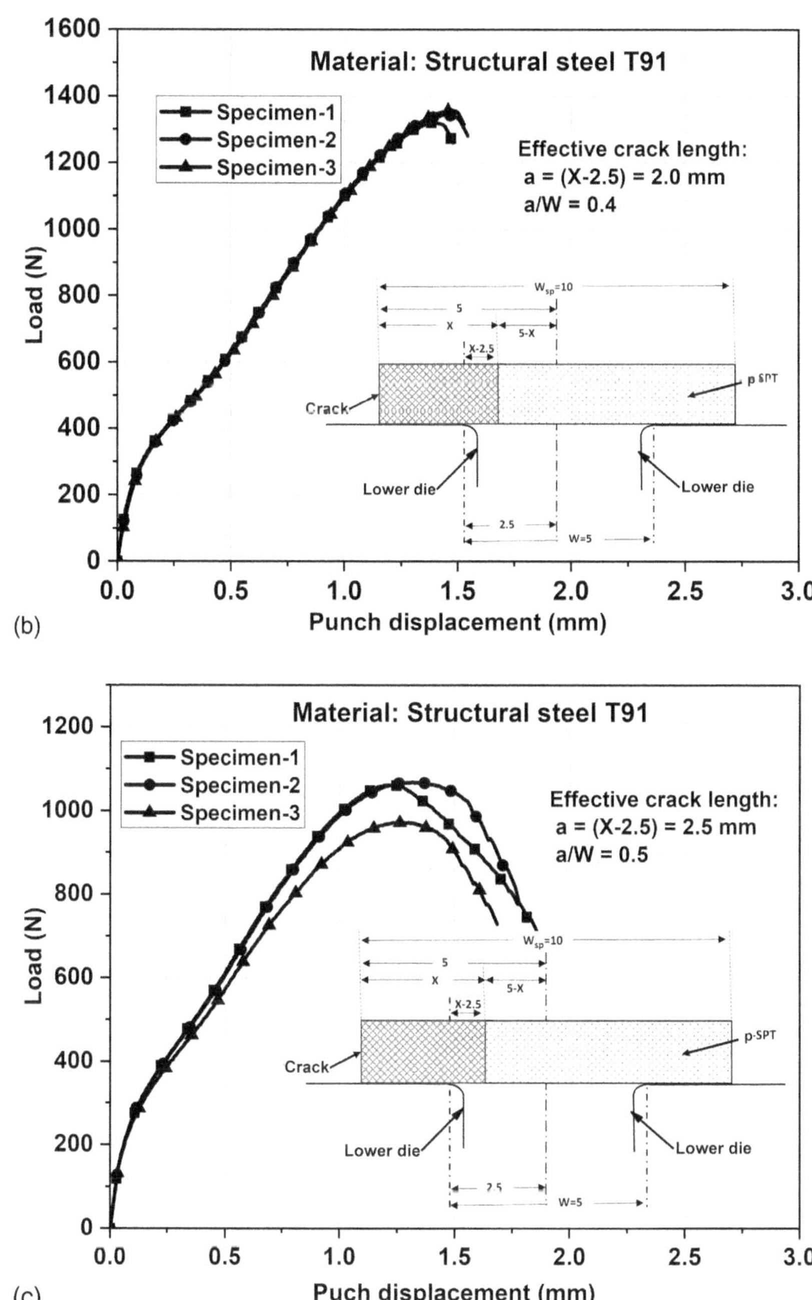

FIG. 7.2 (Continued)

7.2.4 A Parametric Elastic-Plastic FE Study of p-SPT Specimens

As mentioned above, the primary objective of the present chapter is to show the development of a new methodology to calculate the J-R curve of any material using load vs. load-line-displacement data of p-SPT specimens. In this book, this new methodology is called a 'hybrid methodology'. It consists of two steps: (i) the testing of p-SPT specimens with a varying a/w ratio and (ii) the FE elastic-plastic parametric analysis of the p-SPT specimens. The data collected through these two steps are finally used to calculate the J-R data of the material, hence the name 'hybrid methodology'. The previous section dealt with the experimental program, that is, step (i). The present section deals with the FE parametric study of the p-SPT specimens, that is, step (ii), followed by subsequent evaluations.

The quasi-static elastic-plastic FE analyses of the p-SPT specimens are carried out for three values of a/W ratios. Due to symmetricity, half of the pre-cracked specimen is modeled by means of eight-noded fully integrated iso-parametric brick elements. The mesh is refined near the crack tip. The deformations of the punch and dies are considered to be negligible during the analysis due to following two reasons: (a) the dimensions of the p-SPT specimen are of the order of hundreds of microns, whereas the dimensions for the dies and punch are of the order of tens of mm. Hence, the stiffnesses of the punch and dies are many orders higher than that of the p-SPT specimen and (b) the maximum load during the tests is of the order of 1,000 N, which is grossly insufficient to create any noticeable deformations in the dies and punch. Hence, the upper die, lower die, and punch are modeled as rigid bodies in the FE simulations. The multi-body dynamic concept is used to model the progress of the contact surfaces (i) between the specimen and the lower die and also (ii) between the punch and the specimen. In addition, frictional forces are also modeled between these two surfaces. The coefficient of friction is taken as 0.1, as suggested in [9–11]. The von Mises yield criterion, Prandtl–Reuss flow rule, and isotropic strain hardening are used in the elastic-plastic FE analysis. The large displacement and finite strain formulations are employed. The geometric nonlinearity is considered using the updated Lagrangian formulations. The equilibrium state of each load increment is obtained by using the modified Riks algorithm.

Parametric elastic-plastic FE analysis is carried out for six ratios of a/W : 0.3, 0.35, 0.4, 0.45, 0.5, and 0.55. Fig. 7.3 shows the load vs. load line displacement plots obtained through such parametric studies of 20MnMoNi55 and T91 steels. The experimental results shown in Fig. 7.1 and Fig. 7.2, and the results of parametric studies shown in Fig. 7.3 are subsequently used to calculate the material J-R curve in the following section.

7.2.5 Post-processing of the Experimental and FE Results to Calculate the J-R Curve

The experimental and FE analysis data acquired in Sections 7.2.3 and 7.2.4 are then post-processed to calculate the J-R curves of both steels. The steps are shown below.

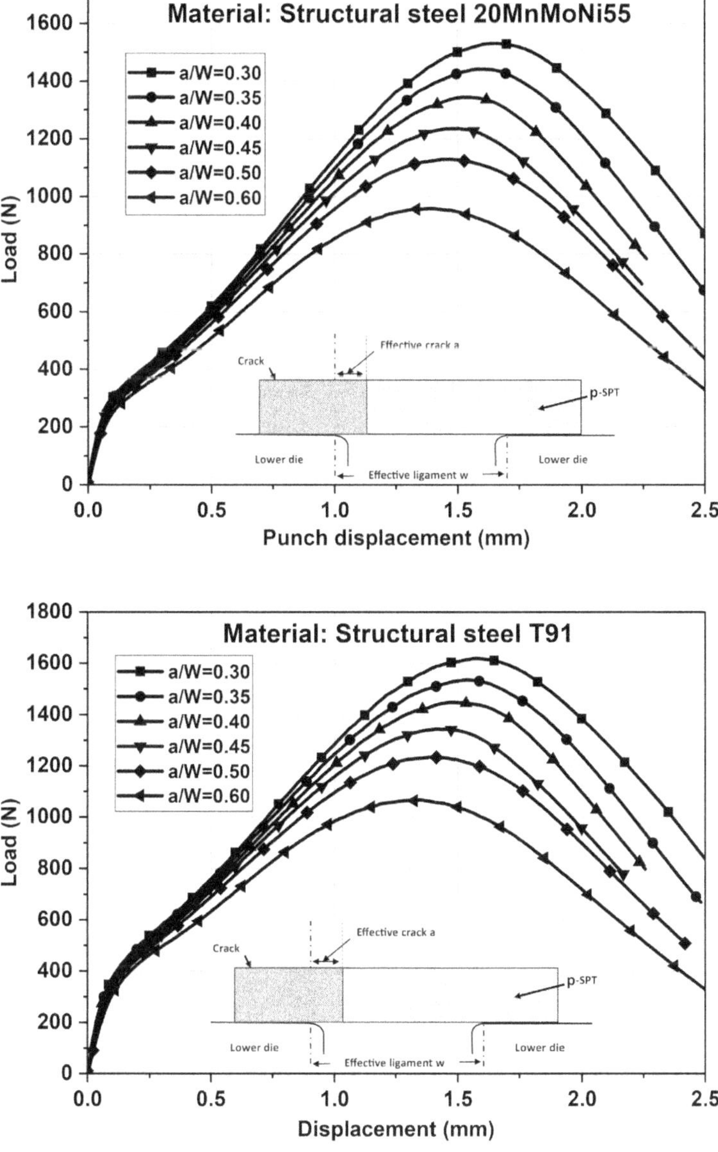

FIG. 7.3 Plots of the parametric FE analysis results of p-SPT specimens for different values of a/W of 20MnMoNi55 and T91 structural steels

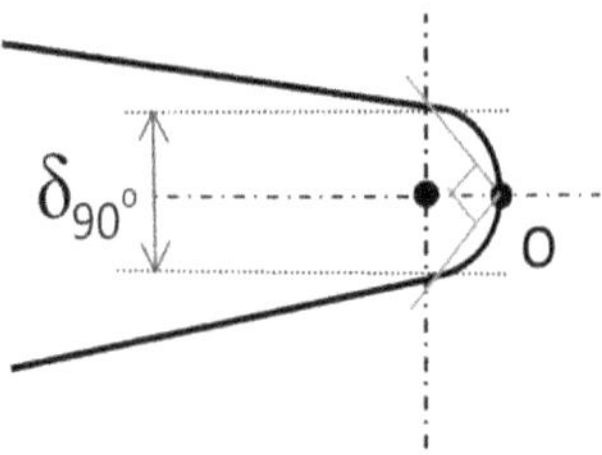

FIG. 7.4 Schematic figure of the CTOD calculated using nodal displacements at the intersection between 90° vertices and the crack flanks

Post-processing of the FE Numerical Results:

a. Using the parametric FE elastic-plastic results shown in Fig. 7.3, the peak loads are plotted as a function of the crack length. The best fit linear equations are derived from these plots for both steels separately. The plots and best fit straight lines are shown in Fig. 7.5 for 20MnMoNi55 and T91 steels. The equations of these straight lines are as follows:

$$a = -0.00257\left(P_L\right) + 5.437 \text{ for the 20MnMoNi55 material} \tag{7.2}$$

$$a = -0.00266\left(P_L\right) + 5.823 \text{ for the T91 material.} \tag{7.3}$$

b. The FE nodal displacements near the crack tip are applied to calculate the crack tip opening displacement (CTOD) at peak loads using the definition of CTOD available in [12]. This is based on the displacement at the intersection between the 90° vertices and the crack flanks, as shown in Fig. 7.4. The CTODs thus calculated are subsequently plotted as a function of the peak loads. The best fit quadratic equations are derived from these plots. The plots are shown in Fig. 7.6 for 20MnMoNi55 and T91 steels. The best fit equations are as follows:

For 20MnMoNi55 material:

$$CTODp = -1.8594 \times 10^{-6}\left(P_L^2\right) + 0.00329\left(P_L\right) - 0.17996 \tag{7.4}$$

For T91 material

$$CTODp = -1.48538 \times 10^{-6}\left(P_L^2\right) + 0.00264\left(P_L\right) + 0.08851. \tag{7.5}$$

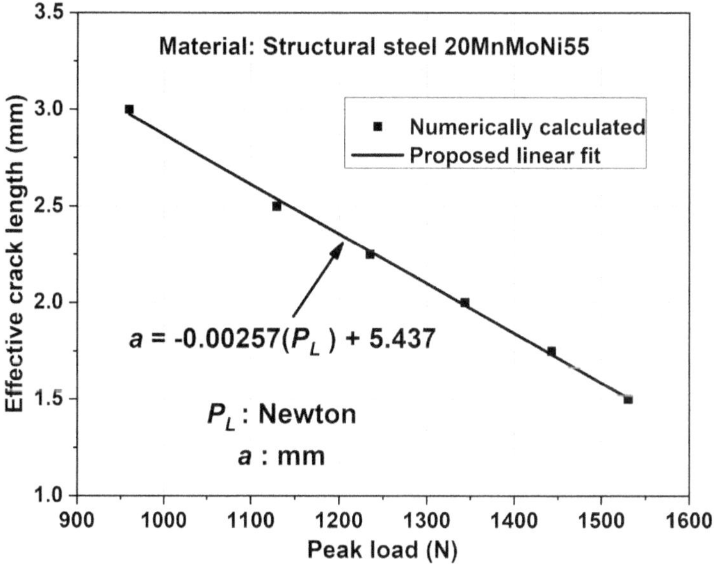

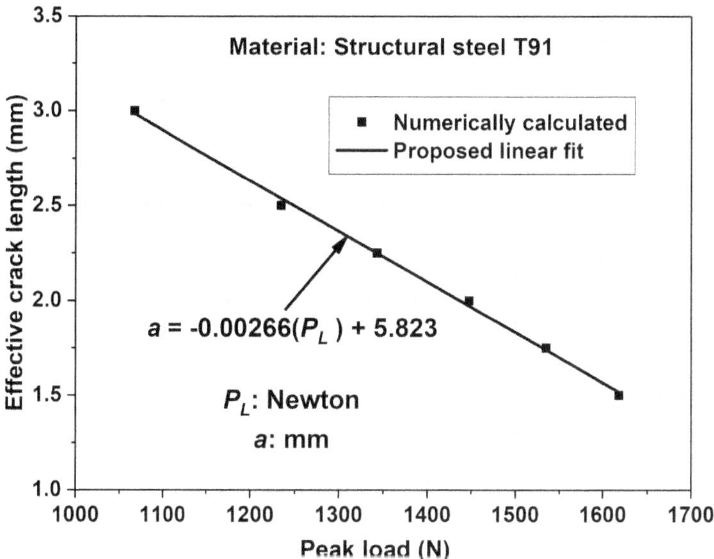

FIG. 7.5 Plots of the crack length vs. peak load of p-SPT specimens of 20MnMoNi55 and T91 structural steels using the FE results of Fig. 7.3

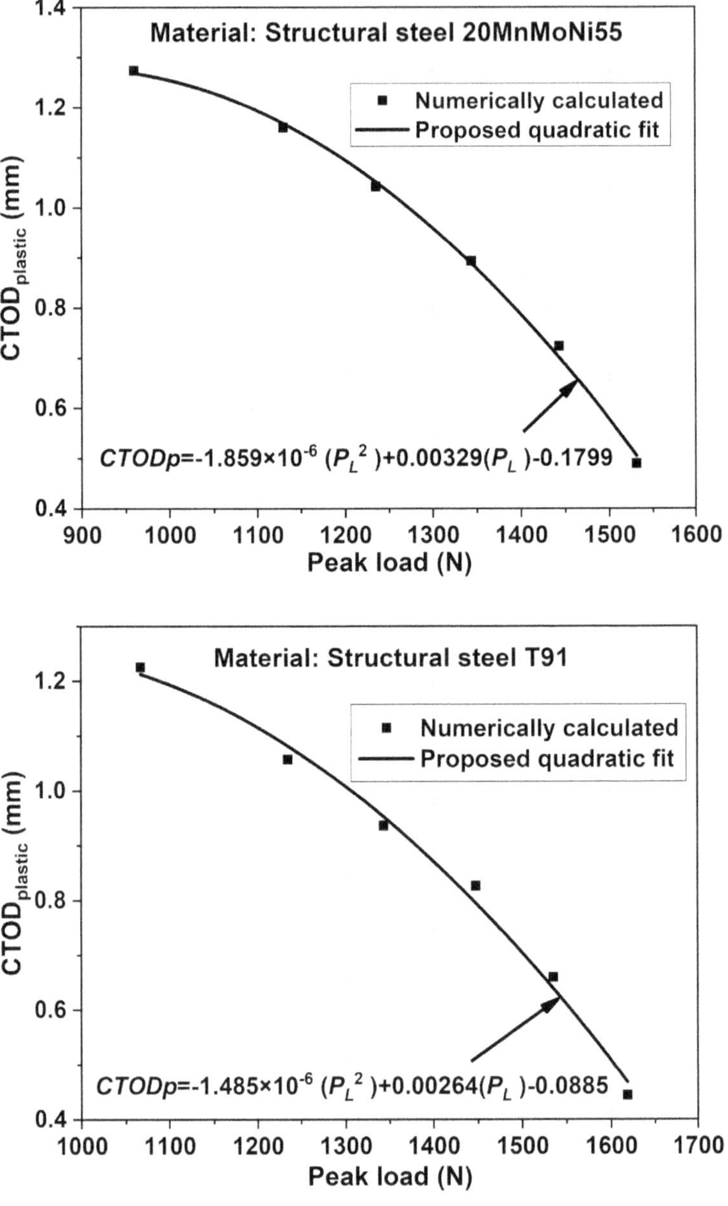

FIG. 7.6 Plots of the plastic CTOD vs. peak load of p-SPT specimens for 20MnMoNi55 and T91 structural steels

Post-processing of Experimental p-SPT Data in Association with eq. (7.2) to eq. (7.5):

After deriving eq. (7.2) to eq. (7.5), experimental data along with these equations are further processed to calculate the J-R curves of both steels as follows:

a. Eq. (7.2) or eq. (7.3) is used to calculate the current effective crack lengths a_p at the peak loads of all the tested specimens. For this purpose, the experimental peak loads shown in Fig. 7.1 or Fig. 7.2 are substituted in eq. (7.2) or eq. (7.3), whichever is applicable.
b. The next task is to calculate crack growth at the peak loads. It is presumed here that for all the tested specimens, the initial crack length is 1.5 mm, which is termed the 'reference effective crack length'. Thus, crack growth Δa_p at the peak load of every specimen is calculated by subtracting 1.5 mm from the estimated crack length at the peak load. This calculation is carried out for all 18 specimens of both steels.
c. Along the same line, eq. (7.4) or eq. (7.5) is used to calculate the CTOD at the experimental peak load CTOD_p of both steels.
d. The data 'crack growth Δa_p vs. peak loads' of all the tested specimens calculated in step b and the data 'CTOD_p vs. peak loads' of all the tested specimens calculated in step c are used to obtain the data 'CTOD_p vs. Δa_p at peak loads' of all the tested specimens. The plots of these data are shown in Figs. 7.7(a) and (b) for 20MnMoNi55 and T91 steels, respectively. Readers may note that the plots CTOD_p vs. Δa_p also represent the crack growth resistance of both steels. This is a direct reflection of the crack growth in the p-SPT specimens during the experiment. Such information can only be obtained reliably by carrying out experiments.
e. The parameter CTOD_p shown in Figs. 7.7 (a) and (b) can then be converted to the J-integral using the expressions of the *m*-factors derived earlier in Chapter 6. Readers may recall that the inputs to calculate the *m*-factors are the values of the effective a/W, the yield stress, and the ultimate stress of the relevant steel. As shown in Chapter 6, there are two *m*-factors for every specimen. These m-factors are valid over the two ranges of the CTOD: m_1 valid over $0 < \text{CTOD} <= 0.1$ and m_2 valid over $0.1 < \text{CTOD}$. These values are calculated for the reference effective $a/W = 0.3$ of both steels. The values are as follows:

$$m_1 = 1.1536486 \text{ and } m_2 = 0.82193387 \text{ for 20MnMoNi steel}$$

$$m_1 = 1.1085933 \text{ and } m_2 = 0.72956294 \text{ for T91 steel.}$$

The procedure described above is shown as a block diagram in Fig. 7.8. This procedure is subsequently applied to calculate the J-R curves of both steels. The J-R curves thus calculated are shown in Fig. 7.9.

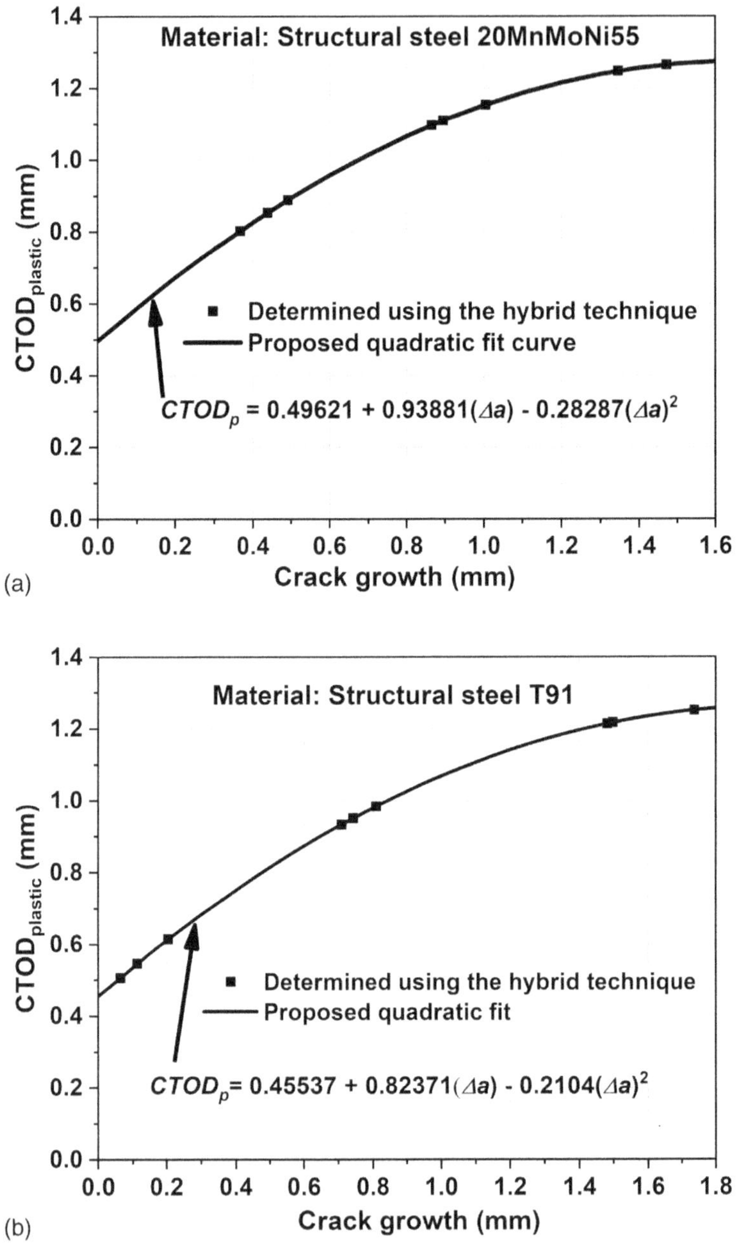

FIG. 7.7 Plots of the calculated CTOD vs. crack growth data for (a) 20MnMoNi55 and (b) T91 structural steels

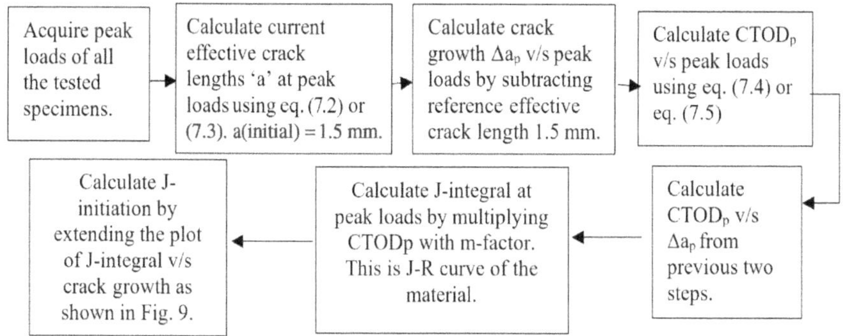

FIG. 7.8 Procedure to calculate the J-R curve applying the new methodology

f. Fig. 7.9 also shows the experimental values of the J-R data taken from the literature on 20MnMoNi55 [13] and T91 [14] steels. In both these references, standard compact tension (CT) specimens with side grooves are used. Such experimental procedure generates the plane strain $J_{initiation}$ and J-R curve. As the thickness of the p-SPT specimens is of the order of 500 microns, the state of stress in the present case is expected to be close to the plane stress condition. Hence, $J_{initiation}$ in the present case is likely to be more due to lower crack tip constraints. This feature is noticeable in both Figs. 7.9 (a) and (b). It may also be seen that the calculated J-R curves of both steels using the present hybrid methodology are in good agreement with the experimental values quoted in [13] and [14] using standard CT specimens.

7.3 PART II: VALIDATION OF THE HYBRID METHODOLOGY USING DAMAGE ANALYSIS OF p-SPT SPECIMENS

Section 7.2 describes a new hybrid methodology to assess the J-R curve of a material using the experimental load vs. punch displacement data of the p-SPT specimen in association with the parametric FE elastic-plastic results. This hybrid methodology is a new methodology and is able to calculate J-R curves very close to the values reported in the literature for two widely used structural steels. In the present section, an independent verification of the new hybrid methodology is further presented to readers. The verification is carried out using the damage mechanics analysis of the p-SPT specimens using the GTN model. The purpose of this section is also to familiarize readers with a procedure to carry out FE analysis of p-SPT specimens using the damage mechanics model to generate the J-R curve of a material.

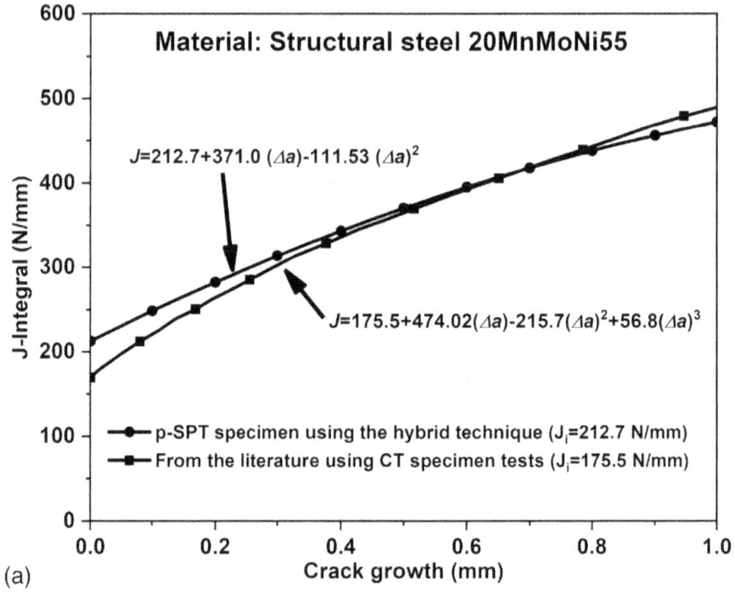

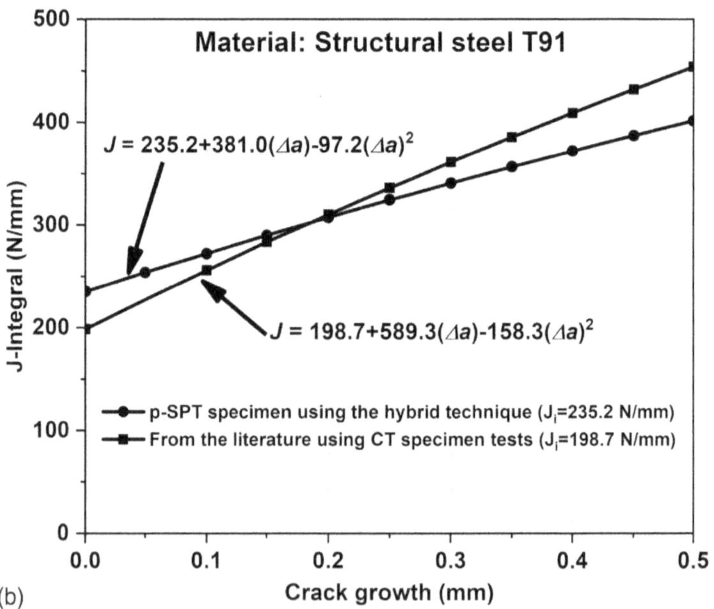

FIG. 7.9 Calculated J-R data of (a) 20MnMoNi55 and (b) T91 steels using the new hybrid methodology in comparison with the literature data

The authors restate that users of the new hybrid methodology need not perform damage mechanics analysis of the p-SPT specimens as part of the procedure described in Section 7.2. The purpose of the present section is only to verify the new methodology so that readers are more informed about its applicability.

7.3.1 FE Analysis using the GTN Model

The GTN model used to carry out the damage analysis of a structure is well covered in the literature. Readers are advised to go through this literature. A brief description of the GTN model is given below. This model is based on the ductile damage processes consisting of the nucleation, growth, and coalescence of micro-voids. The yield function introduced by Gurson [15] and later modified by Tvergaard and Needleman [16] is

$$\varnothing\left(\sigma_e,\sigma_h,\sigma_y,f\right)=\left(\frac{\sigma_e}{\sigma_y}\right)^2+2f^*q_1 cosh\left\{\frac{3q_2\sigma_h}{2\sigma_y}\right\}-\left\{1+q_3\left(f^*\right)^2\right\}=0, \tag{7.6}$$

where f is the micro-void volume fraction, σ_h is the hydrostatic stress, σ_e is the von Mises equivalent stress, σ_y is the flow stress of the material matrix, and q_1, q_2, q_3 are the fitting parameters suggested by Tvergaard [16, 17]. The yield function above is transformed to von Mises yield criteria when the void volume fraction f^* is zero. The modified void volume fraction f^* was also introduced by Tvergaard and Needleman [18] to predict the rapid loss in material strength after the commencement of void coalescence. This is expressed as follows:

$$f^* = \begin{cases} f & \text{for } f \le f_c \\ f_c + \dfrac{f_u^* - f_c}{f_f - f_c}\left(f - f_c\right) & \text{for } f > f_c \end{cases} \tag{7.7}$$

where f_c is the critical void volume fraction, f_f is the void volume fraction at fracture, and $f_u^* = 1/q_1$ is the ultimate void volume fraction. The function becomes more predominant once the void volume fraction f exceeds a critical coalescence value f_c.

The evolution law of the void volume fraction is given in the model as follows:

$$\dot{f} = \dot{f}_{growth} + \dot{f}_{nucleation}, \tag{7.8}$$

where $\dot{f}_{growth}$ is the change in void volume fraction due to void growth and $\dot{f}_{nucleation}$ is due to the nucleation of new voids. According to Chu and Needleman [19], the nucleation rate is assumed to follow a Gaussian distribution, expressed as

$$\dot{f}_{nucleation} = A\dot{\bar{\varepsilon}}^{P}, \tag{7.9}$$

where $\dot{\bar{\varepsilon}}^{P}$ is the equivalent plastic strain rate and

$$A = \frac{f_n}{S_n\sqrt{2\pi}}\exp\left(-\frac{1}{2}\left(\frac{\bar{\varepsilon}^{P}-\varepsilon_n}{S_n}\right)^2\right), \tag{7.10}$$

where ε_n is the mean strain, S_n is the standard deviation, and f_n is the void volume fraction at nucleation.

7.3.2 Determination of the GTN Material Parameters

One of the important tasks when making use of the GTN model is to determine the GTN material damage parameters. There are nine such material parameters. Experimental procedures are available in the literature to determine such parameters. However, such experimental procedures are complex, require advanced experimental techniques, and involve uncertainties. Such material parameters can also be determined numerically by carrying out parametric studies and comparing the computed load vs. displacement data with the experimental results. It has been noticed that different GTN parameters have distinct effects on the various sections of a load vs. displacement plot. This aspect can be utilized to ascertain the GTN parameters by carrying out a parametric study of the GTN parameters and carefully comparing the computed data with the experimental results of different sections of the load vs. displacement curve. In principle, such a methodology can be used to determine a unique set of GTN parameters applicable to the given material. However, this is an involved task and susceptible to errors.

In the present section, readers can become familiar with an advanced methodology to determine the GTN parameters numerically using an artificial neural network (ANN). To carry out a parametric study, a range of GTN parameters are selected such that all experimental results lie within the bounds of the simulation results. The computed load vs. displacement data, along with the respective GTN parameters, are then applied to train an ANN. The trained network is subsequently used to establish the actual GTN material parameters using the experimental load vs. displacement data as input to the ANN. It may be noted that the ANN helps to identify a unique set of GTN parameters by minimizing the residue. This methodology is described in the following sections through case studies involving p-SPT specimens, for which experimental data are presented in Section 7.2.

7.3.3 FE Modeling of a p-SPT Specimen using the GTN Damage Model: Parametric Studies

a. Three-dimensional symmetrical FE models of p-SPT specimens described in Section 7.2.3 are analyzed using the GTN damage model. Some important features of the FE model, which may help readers to carry out similar GTN analyses of p-SPT specimens, are as follows: The minimum length of the brick element is taken as 35.7 μm along the crack line and also in the thickness direction. This mesh size near the crack tip is chosen so that the computed peak load matches well with the peak load measured during the experiment. The upper die, lower die, and punch are modeled as rigid bodies. The friction factor between the surface of the indenter and the p-SPT specimen is taken as 0.1, as explained earlier. Displacement-controlled loading is applied to the rigid punch.
b. As mentioned above, the GTN model requires nine material parameters. In the present case study, the values of the five parameters, that is, q_1, q_2, q_3, ε_n, and S_n are taken from the literature [16, 19]. In addition, the value of f_0 is assumed to be negligible as the inclusion ratings of the present steels are very low. The GTN parameters taken from the literature are shown in Table 7.3. These values are kept constant during the parametric study.
c. To ascertain the remaining three GTN parameters, that is, void volume fraction at nucleation (f_n), void volume fraction at coalescence (f_c), and void volume fraction at fracture (f_f), an FE parametric simulation of a p-SPT specimen with a/W equal to 0.50 is carried out. To carry out a parametric study, variations in f_n, f_c, and f_f are chosen to have the experimental load vs. punch displacement data lie within the bounds of the computed results, as shown in Fig. 7.10. With this consideration, it is found that, for the present case study, the sets of GTN parameters of 20MnMoNi55 steel lie within the ranges $0.01 \le f_n \le 0.015$, $0.015 \le f_c \le 0.025$, and $0.2 \le f_n \le 0.3$. Similarly, the GTN parameters of T91 steel lie within the ranges $0.01 \le f_n \le 0.015$, $0.015 \le f_c \le 0.025$, and $0.15 \le f_n \le 0.25$. Fig. 7.10 shows the computed load vs. punch displacement plots of both steels for different sets of GTN parameters. The GTN parameters in each case and the experimental results are also shown in these figures.
d. An ANN is then trained using the computed data shown in Fig. 7.10. The ANN applied in the present case study is shown in Fig. 7.11. It consists of (i) an input layer that has 11 neurons to input 11 computed loads at pre-defined

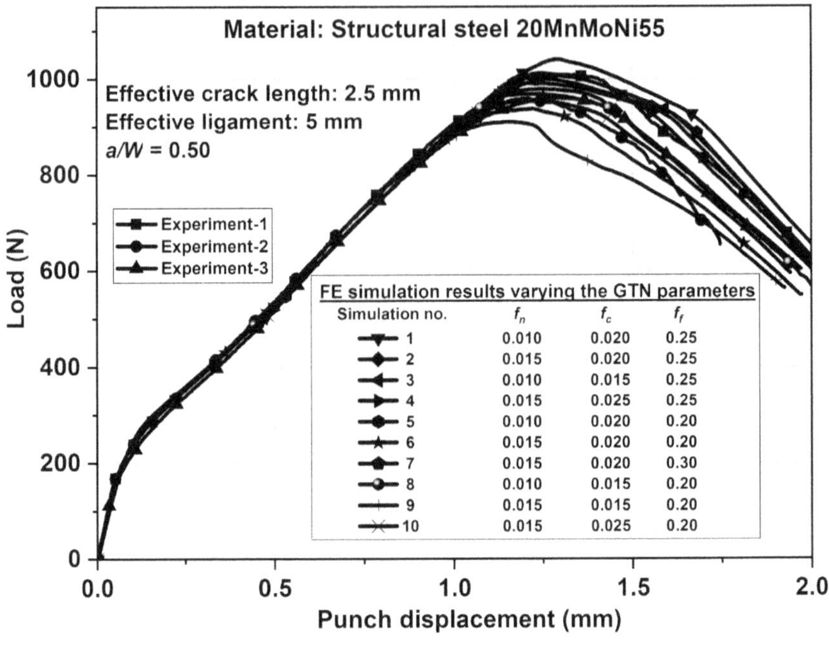

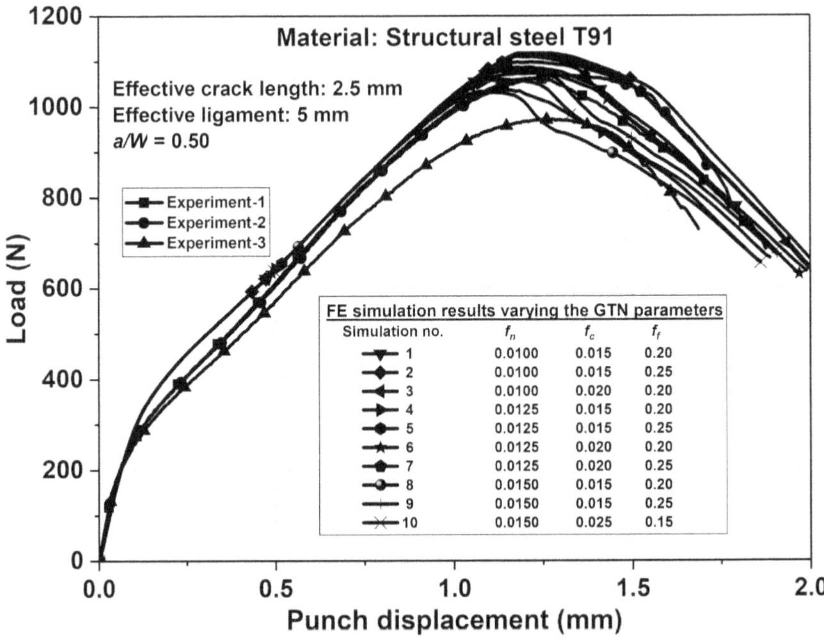

FIG. 7.10 Plots of the load vs. load line displacement data of p-SPT specimens with $a/W = 0.50$ for different sets of GTN parameters of 20MnMoNi55 and T91 steels

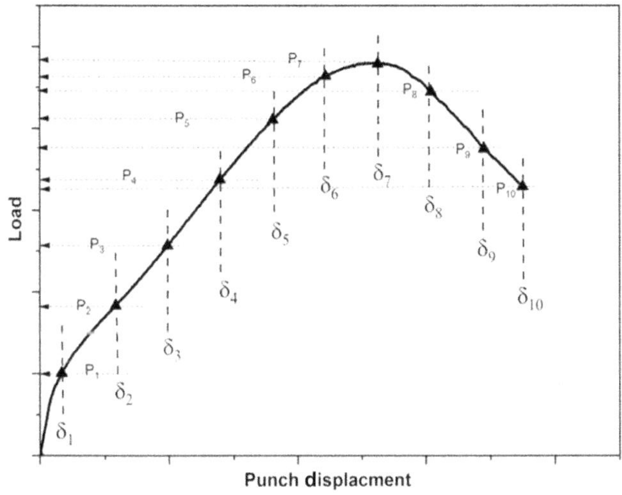

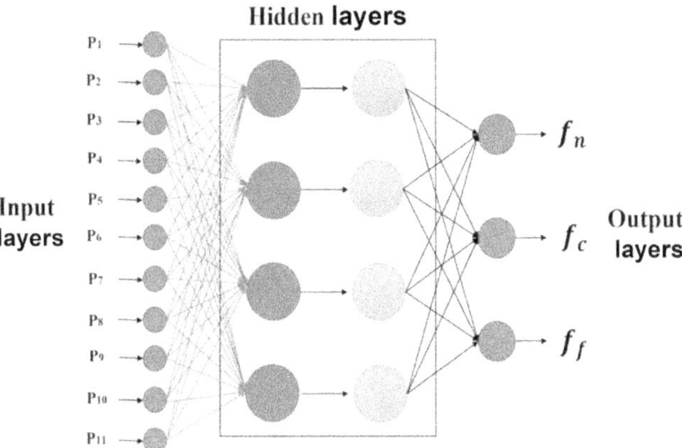

FIG. 7.11 Schematic diagram of the ANN and corresponding neuron inputs, and outputs applied in the present study

TABLE 7.3
Values of six GTN parameters taken from the literature, which are kept constant during the analysis

q_1	q_2	q_3	ε_n	S_n	f_0
1.5	1.0	2.25	0.3	0.1	1.0^{-5}

TABLE 7.4
Calculated GTN parameters applying a trained ANN for 20MnMoNi55 steel

20MnMoNi55 structural steel	Specimen a/W	Void volume fraction at nucleation (f_n)	Void volume fraction at coalescence (f_c)	Void volume fraction at fracture (f_f)
Specimen-1	0.5	0.01189	0.01965	0.2492
Specimen-2	0.5	0.01371	0.02141	0.2003
Specimen-3	0.5	0.01296	0.02104	0.2258
Average	-	0.01285	0.02070	0.2251

TABLE 7.5
Calculated GTN parameters applying trained ANN for T91 steel

T91 structural steel	Specimen a/W	Void volume fraction at nucleation(f_n)	Void volume fraction at coalescence (f_c)	Void volume fraction at fracture (f_f)
Specimen-1	0.5	0.01314	0.01909	0.2035
Specimen-2	0.5	0.01055	0.02021	0.1963
Average	-	0.01185	0.01965	0.1999

punch displacements, as shown in Fig. 7.11, (ii) two hidden layers, where each consists of four neurons each, and (iii) an output layer that consists of three neurons representing three GTN parameters: f_n, f_c, and f_f. The ANN is trained with the data shown in Fig. 7.10 for each steel separately. The trained ANN is then applied to calculate the GTN parameters f_n, f_c, and f_f of each steel by inputting experimental loads at the same pre-defined punch displacements as used during training. The GTN parameters determined by such a procedure are shown in Table 7.4 and Table 7.5 for each tested specimen of both steels . The average values are taken as applicable GTN parameters for each steel for subsequent analysis.

e. The average GTN parameters shown in Table 7.4 and Table 7.5 are then applied to carry out damage mechanics analysis of the p-SPT specimens. The experimental results of these p-SPT specimens are reported earlier in Section 7.2.3. Fig. 7.12 and Fig. 7.13 show the comparison of the computed load vs. punch displacement data with the experimental results of 20MnMoNi55 and T91 steels, respectively. It may be noted that the computed load vs. punch displacement data are in good agreement with the experimental results while using GTN parameters calculated by applying an ANN. This indicates that the application of an ANN to determine material GTN parameters is a valuable tool.

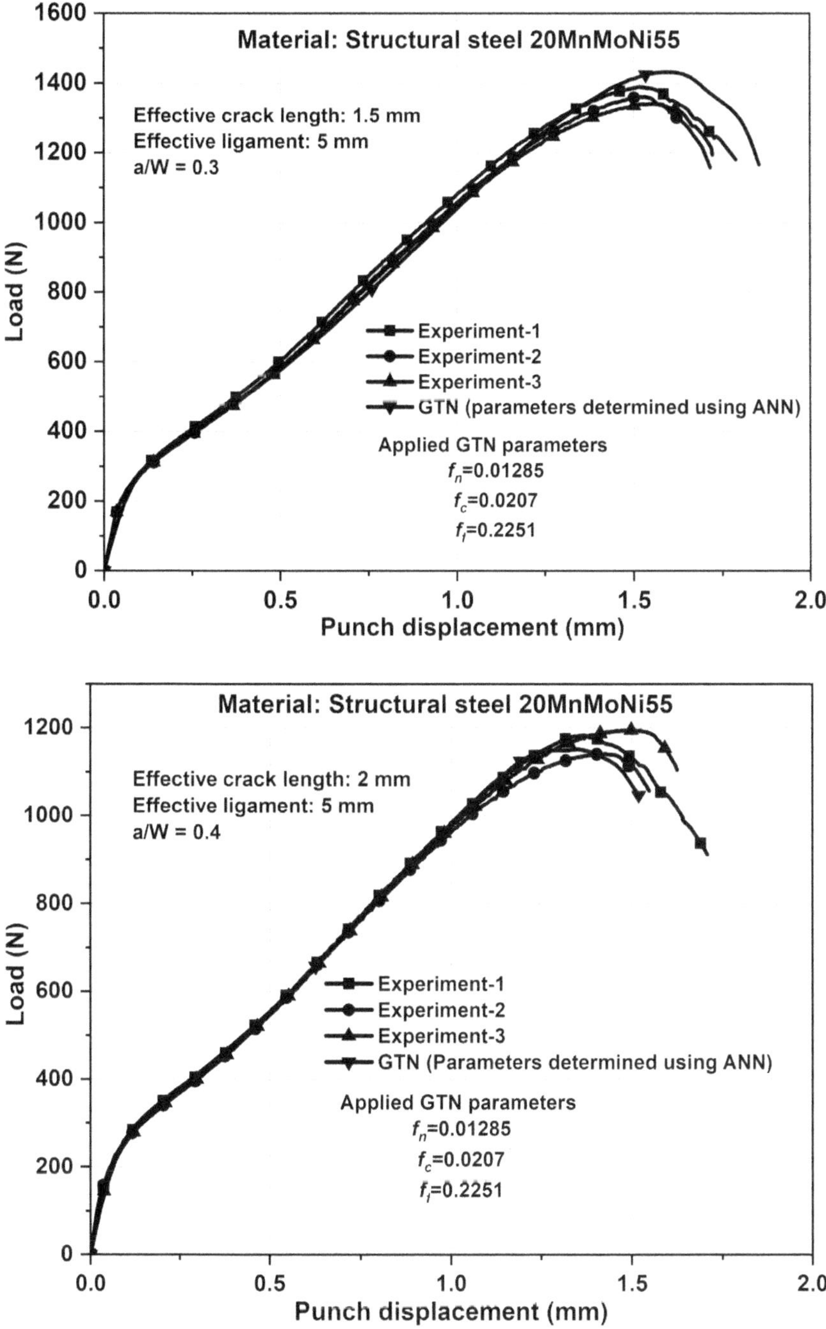

FIG. 7.12 Plots of the load vs. punch displacement of p-SPT specimens calculated using the GTN model in comparison with the experimental data of 20MnMoNi55 steel for three crack lengths

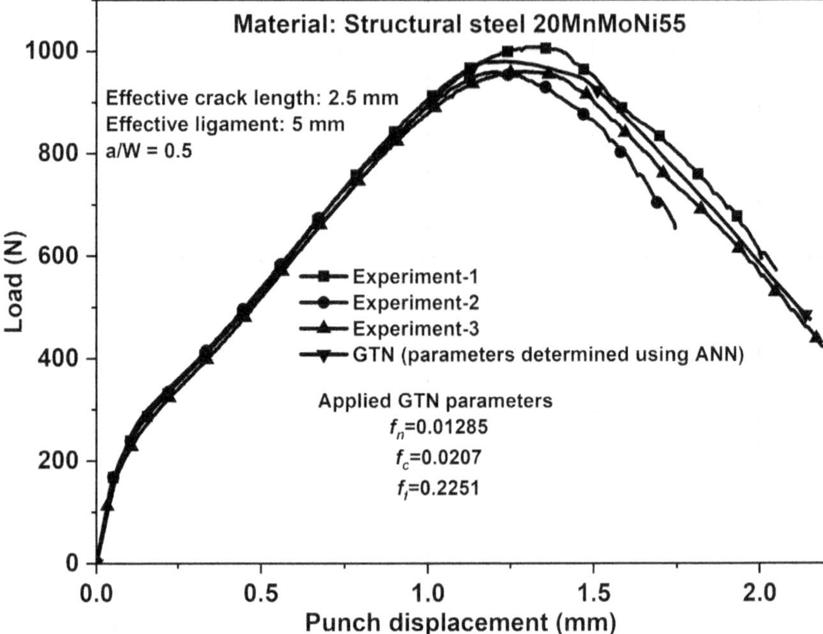

FIG. 7.12 (Continued)

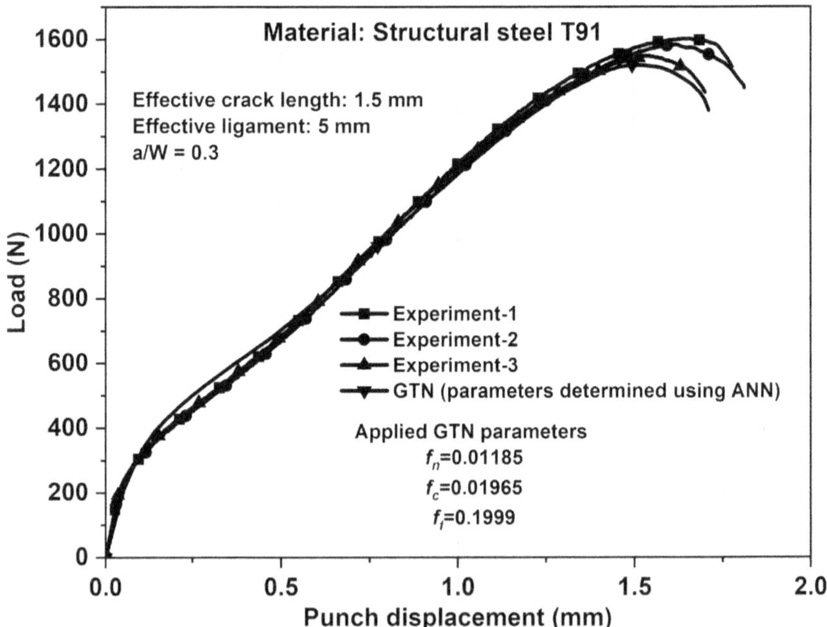

FIG. 7.13 Plots of the load vs. punch displacement of p-SPT specimens calculated using the GTN model in comparison with the experimental data of T91 steel for three crack lengths

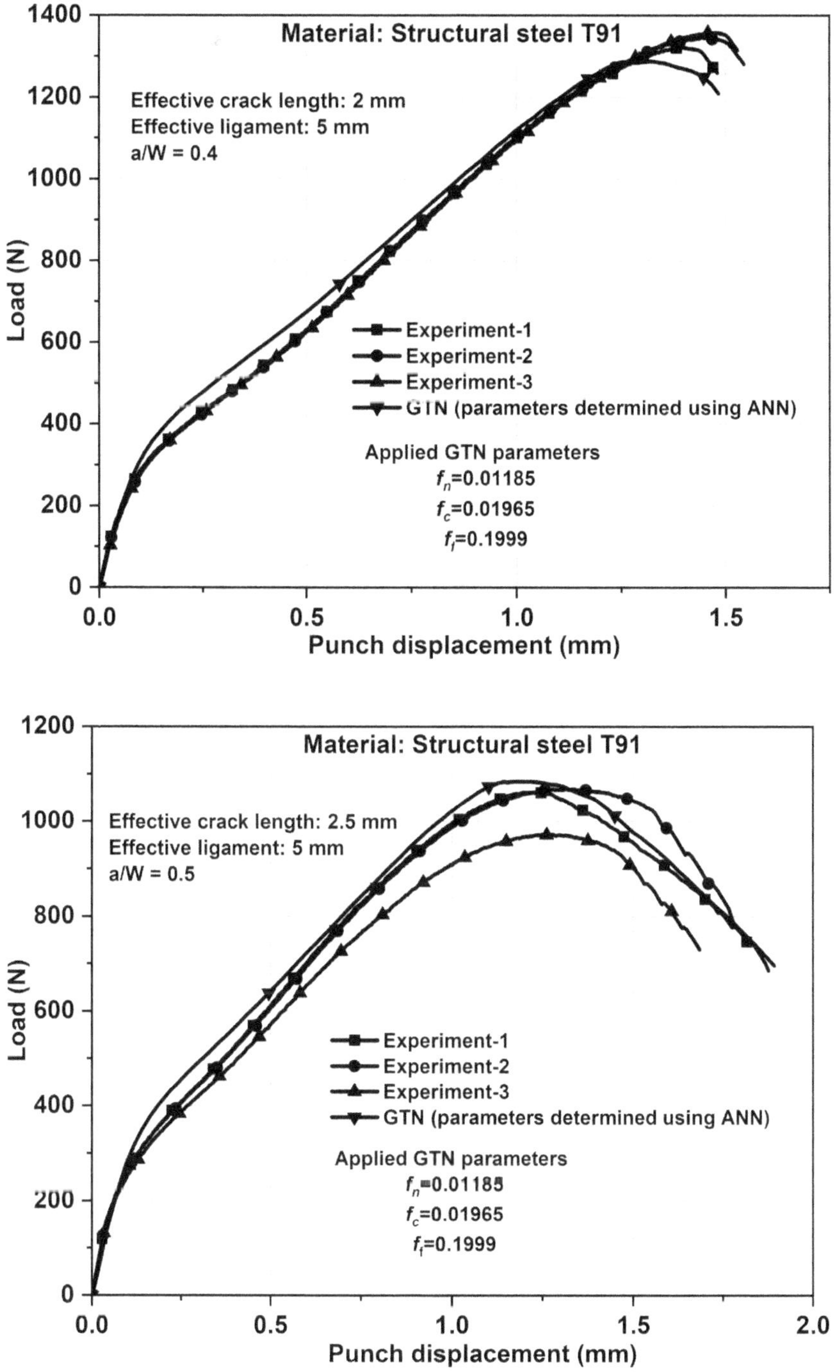

Material: Structural steel T91
Effective crack length: 2 mm
Effective ligament: 5 mm
a/W = 0.4
Load (N)
Punch displacement (mm)
Experiment-1
Experiment-2
Experiment-3
GTN (parameters determined using ANN)
Applied GTN parameters
f_n=0.01185
f_c=0.01965
f_f=0.1999

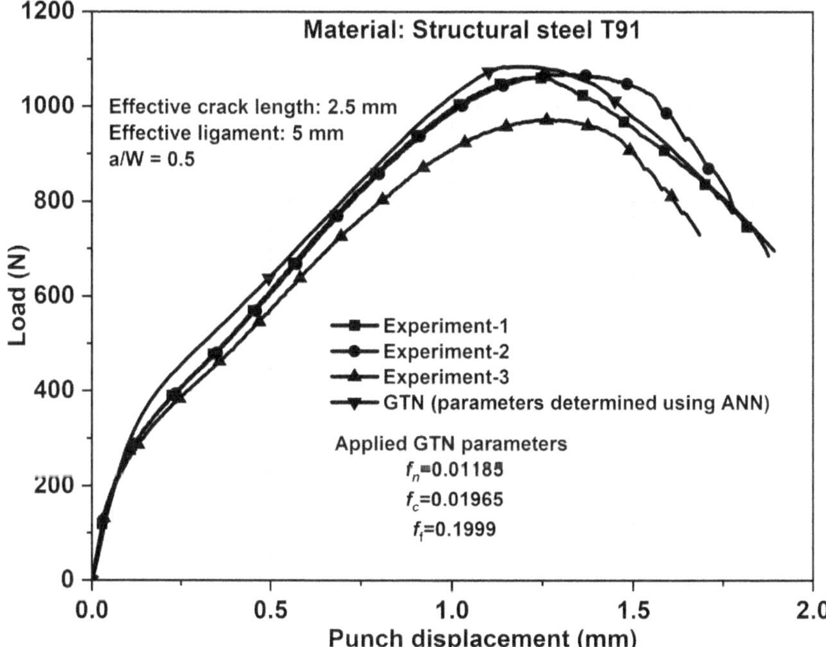

FIG. 7.13 (Continued)

Readers may note that the load vs. punch displacement data shown in Figs. 7.12 and 7.13 are direct outputs of the FE computation with damage mechanics. However, to determine the fracture parameters, one needs to post-process the FE results. Such fracture parameters are: (i) the crack growth, (ii) CTOD, and (iii) J-integral as a function of the load increments. The data on CTOD vs. crack growth or J-integral vs. crack growth represents the material resistance against crack growth. In the following subsections, methodologies to calculate these parameters using the results of the GTN model are explained. Subsequently, such parameters are also compared with the results calculated by new hybrid methodology as presented in Section 7.2. A close comparison confirms the usefulness of the hybrid methodology.

f. Calculation of crack growth using the results of the GTN model: Readers may note that the analysis carried out by employing the GTN model of any cracked specimen implicitly models the process of crack growth. With the load increment, the Gauss points in the FE model close to the crack tip undergo damage, quantified by the void volume fraction. A Gauss point is assumed to be completely damaged when its void volume fraction attains the void volume fraction at fracture f_f, which is input as one of the GTN material parameters. A completely damaged Gauss point implies crack growth up to such a location of the Gauss point in the specimen. This inference is applied to calculate crack growth in the p-SPT specimens. This is done by monitoring the sequence of Gauss points that undergo complete damage near the crack tip during the increment of punch displacement. Fig. 7.14 shows plots of a computed crack length at peak load vs. peak load for 20MnMoNi55 and T91 materials as calculated using GTN analysis. For the sake of comparison, Figs. 7.14 (a) and (b) also show similar plots calculated using the hybrid methodology presented in Fig. 7.5. The equations of the best fit straight lines of these plots are also shown in Figs. 7.14 (a) and (b). These plots indicate the ability of the very simplified new hybrid methodology to calculate such data in close comparison with the data calculated by the GTN model.
g. Calculation of CTOD using the results of the GTN model: The definition of CTOD was introduced in Section 7.2.5 for a stationary crack. However, one needs to update this definition for a growing crack. As such, the definition of CTOD with 90° intercepts loses its physical meaning for a growing crack tip. Readers are advised to go through the voluminous literature available on the procedure to calculate the CTOD in a fracture specimen. The CTOD in the present case studies is calculated with 90° intercepts at the 'original crack tip' [20] to consider the effect of a growing crack. This definition of CTOD of a growing crack tip is illustrated in Fig. 7.15. This definition of CTOD is employed in the present case study to take into effect crack growth due to damage mechanics modeling of the p-SPT specimen. The calculated CTOD as a function of punch displacement for different values of a/W are shown in Figs. 7.15(a) and (b) for 20MnMoNi55 and T91 steels.
h. Calculation of the J-integral using the results of the GTN model: As mentioned earlier, the J-R curve is a measure of material resistance against crack growth. In the present case, the variations of the CTOD shown on the y-scale of Figs. 7.15

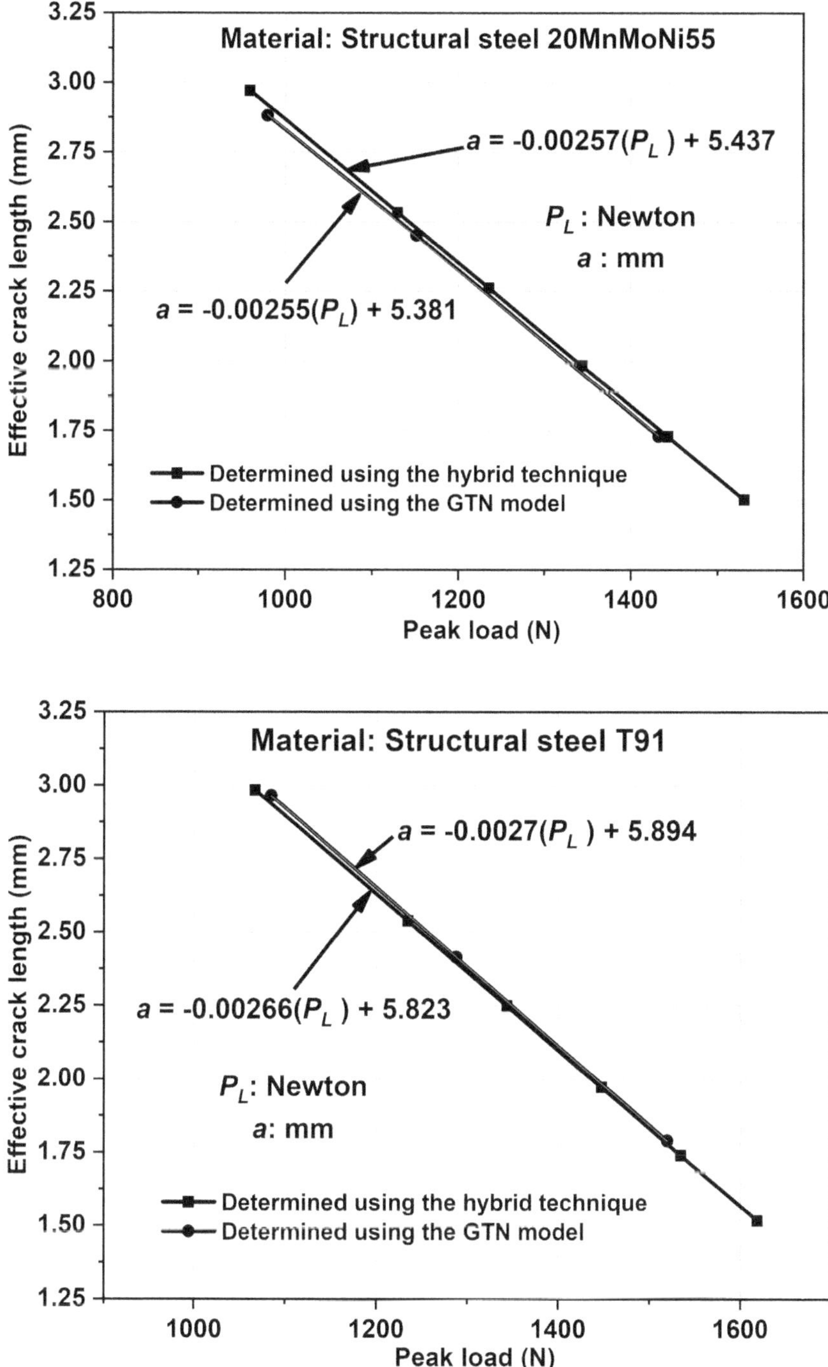

FIG. 7.14 Comparison of peak load vs. current crack length at peak load calculated using the new hybrid methodology and GTN model of 20MnMoNi55 and T91 steels

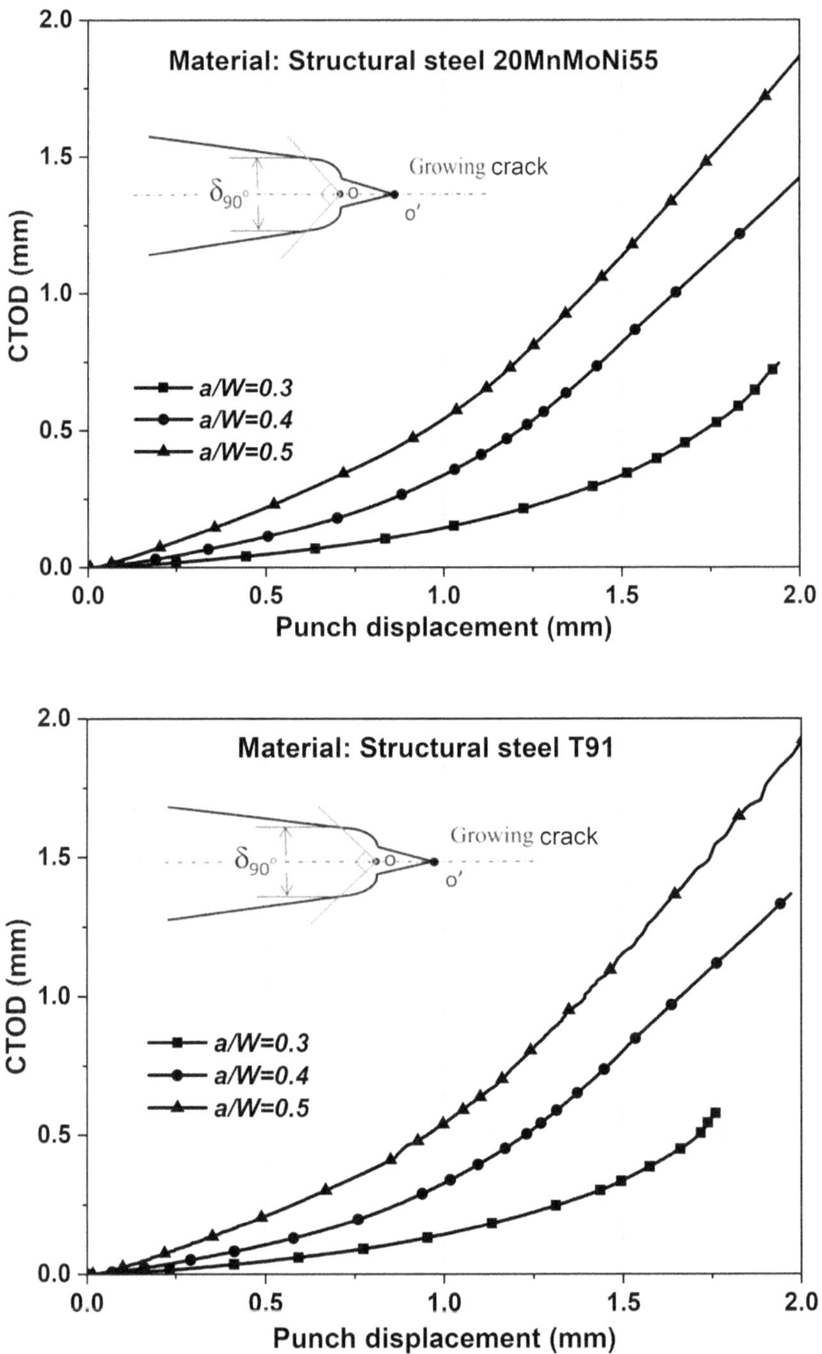

FIG. 7.15 Plots of the CTOD vs. punch displacement of p-SPT specimens using the GTN model of 20MnMoNi55 and T91 structural steels

TABLE 7.6
Comparison of the J-initiations of 20MnMoNi55 and T91 steels calculated using (i) the new hybrid methodology, (ii) FE GTN damage model, and (iii) experimental values quoted in the literature by testing CT specimens

Material	Estimated J-initiation (N/mm) using p-SPT specimens		J-initiation (N/mm) quoted in the literature by testing CT specimens
	Estimated J_i using the new hybrid methodology	Estimated J_i using the GTN damage model	
20MnMoNi55	212.7	199.5	175.5
T91	235.2	234.8	198.7

(a) and (b) as a function of punch displacement are converted to the J-integral adopting the procedure shown in Section 7.2.5 (d). In addition, the procedure to calculate crack growth using the FE GTN model is explained in Section 7.3.3 (f). The same procedure is applied here to calculate crack growth Δa as a function of punch displacements. A plot between J-integral and crack growth Δa represents the J-R curve of the material. The J-R curves of both steels are shown in Figs. 7.16 (a) and (b) as calculated using the above procedure. Readers may note the zigzag pattern of the J-R curves in these figures. This is due to the discrete increase in crack growth with the load increments when carrying out damage mechanics analysis.

Table 7.6 shows a comparison of J-initiation (a) as calculated using the new hybrid methodology introduced in Part I of the present chapter and (b) as calculated using the FE GTN model, and (c) the measured values using CT specimens taken from the literature. It may be noted that the J-initiations calculated by methodologies (a) and (b) are higher than the experimental values (c). Such differences are expected due to lower stress constraints at the crack tip of the p-SPT specimen in comparison to the ASTM standard CT specimens. In general, the J-R plots determined by all the three methodologies are in good agreement.

7.4 CHAPTER CLOSURE

There are three methodologies to calculate the J-R curve of a material:

1. The first method is to test ASTM standard CT specimens to obtain the J-R data of the material. Voluminous information is available in the literature on such an experimental method. The primary limitation of this method is the requirement of a substantial quantity of material to prepare CT specimens of the required

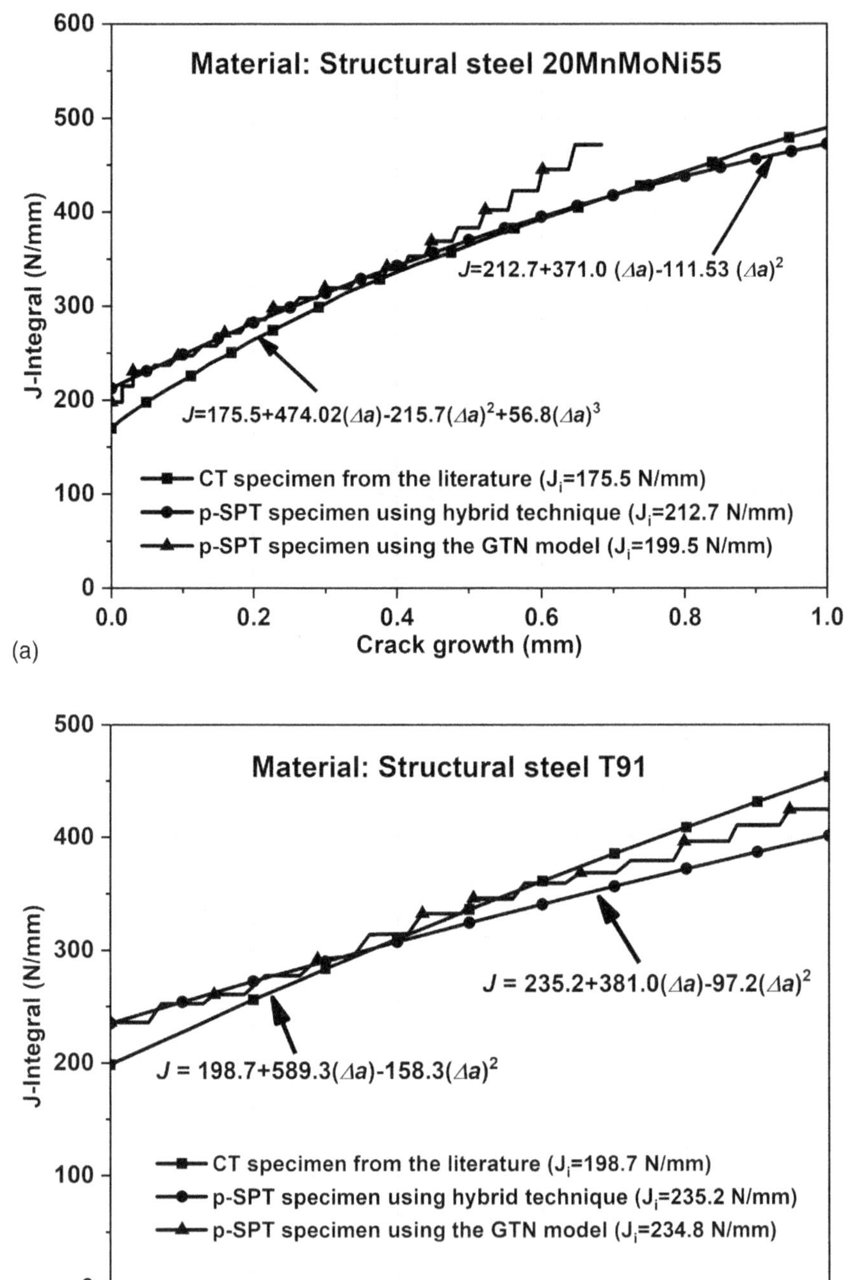

FIG. 7.16 Comparison of J-R data calculated using the three methodologies: the GTN model, present hybrid methodology, and experimental data of (a) 20MnMoNi55 and (b) T91steels

thickness to obtain the desirable degree of stress triaxiality. Such a thickness requirement is significantly more in the case of materials with high ductility.
2. The second method involves the application of the FE method with the GTN model. In this method also, one needs the experimental data to calculate the GTN material parameters (f_n, f_c, and f_f). The GTN material parameters can also be found using the p-SPT experimental data and employing an ANN, as explained in Part II of this chapter. Having calculated the GTN parameters, one can perform GTN damage analysis of the p-SPT specimens to obtain the J-R data. The determination of the GTN material parameters and handling of advanced FE code for damage mechanics analysis require expertise and are subject to computational uncertainties.
3. The third method, termed the 'hybrid methodology' in this chapter, is a newly proposed method. In this method, the experimental p-SPT load vs. punch displacement data in association with the standard elastic-plastic FE analysis of p-SPT specimens are applied to calculate the J-R curve. This method is explained in Part I of this chapter.

The primary objective of this chapter is to acquaint readers with this new methodology to calculate the material J-R curve in the case of the availability of a very limited quantity of material or voluminous material that is unsafe to handle. Some of the examples are aged materials from chemical plants, bio-organic plants, and nuclear plants. In such cases, the new method is very useful. On the one hand, a large quantity of material is not required by this methodology to carry out tests, and on the other hand, this methodology does not involve advanced FE analysis with intricate material parameters, such as GTN parameters. The present chapter also deals with the validation of the new hybrid methodology. This methodology can also be applied, for example, to materials at different temperatures and in irradiated conditions, provided (a) material stress-strain data are known to carry out conventional elastic-plastic FE analysis and (b) experimental p-SPT data are available for different initial crack lengths.

REFERENCES

[1] CWA 15627, Small Punch Test Method for Metallic Materials: Part A and Part B. CEN WS21 (2007)
[2] Kumar P., Dutta B. K., Chattopadhyay J., Fracture toughness prediction of reactor grade materials using pre-notched small punch test specimens. Journal of Nuclear Materials 495 (2017) 351 362.
[3] Alegre J. M., Cuesta I. I., Barbachano H. L., Determination of the fracture properties of metallic materials using pre-cracked small punch tests. Fatigue and Fracture of Engineering Materials and Structures (2015) 38 104–112.
[4] Cuesta I. I., Rodriquez C., Belzunce F. J., Alegre J. M., Analysis of different techniques for obtaining pre-cracked/notched small punch test specimens. Engineering Failure Analysis 18 (2011) 2282–2287.

[5] Alegre J. M., Lacalle R., Cuesta I. I., Álvarez J. A., Different methodologies to obtain the fracture properties of metallic materials using pre-notched small punch test specimens. Theoretical and Applied Fracture Mechanics 86 (2016) 11–18.
[6] Cuesta I. I., Alegre J. M., Influence of biaxial pre-deformation on fracture properties using pre-notched small punch specimens. Engineering Fracture Mechanics 131 (2014) 1–8.
[7] Cuesta I. I., Alegre J. M., Determination of the fracture toughness by applying a structural integrity approach to pre-cracked Small Punch Test specimens. Engineering Fracture Mechanics 78 (2011) 289–300.
[8] Cuesta I. I., Alegre J. M., Determination of plastic collapse load of pre-cracked Small Punch Test specimens by means of response surfaces. Engineering Failure Analysis 23 (2012) 1–9.
[9] Martínez-Pañeda E., Cuesta I. I., Peñuelas I., Díaz A., Alegre J. M., Damage modeling in Small Punch Test specimens. Theoretical and Applied Fracture Mechanics 86 (2016) 51–60.
[10] García T. E., Rodríguez C., Belzunce F. J., Cuesta I. I., Development of a new methodology for estimating the CTOD of structural steels using the small punch test. Engineering Failure Analysis 50 (2015) 88–99.
[11] Martínez-Pañeda E., García T. E., Rodríguez C., Fracture toughness characterization through notched small punch test specimens. Materials Science and Engineering A 657 (2016) 422–430.
[12] Rice J. R., A path independent integral and the approximate analysis of strain concentration by notches and cracks. Journal of Applied Mechanics 35 (1968) 379–386.
[13] Blauel J. G., Hodulak L., Hollstein T., Voss B., Material characterization by J-R curves of a 20MnMoNi55 forging. International Journal of Pressure Vessels and Piping, 17 (1984) 139–162.
[14] Chaouadi R., Flaw and fracture behavior of 9 % Cr-ferritic / martensitic steels. RESTRICTED Contract Report No. SCK•CEN-R-4122 , SCK-CEN, Boeretang, Belgium (2005).
[15] Gurson A.L., Continuum theory of ductile rupture by void nucleation and growth: Part-I Yield criteria and flow rules for porous ductile media. Journal of Engineering Materials and Technology (1977) 99(1) 2–15.
[16] Tvergaard V., On localization in ductile materials containing spherical voids. International Journal of Fracture (1982)18 237–252.
[17] Tvergaard V., Influence of voids on shear band instabilities under plane strain conditions. International Journal of Fracture (1981) 17 389–407.
[18] Tvergaard V., Needleman A., Analysis of the cup-cone fracture in a round tensile bar. Acta Metallurgica (1984) 32 157–169.
[19] Chu C. C., Needleman A., Void nucleation effects in biaxially stretched sheets. Journal of Engineering Materials and Technology (1980) 102 249–256.
[20] Zhu X., Zelenak P., Mcgaughy T., Comparative study of CTOD-resistance curve test methods for SENT specimens. Engineering Fracture Mechanics (2017) 172 17–38.

8 Determination of the Mechanical Properties of Degraded Engineering Materials using SPT Data

A Real-Life Case Study

S.R. Ghodke and B.K. Dutta

8.1 OVERVIEW

One of the major factors that may impact the extended service life of plant components is material degradation. The assessment of material degradation, estimation of the remaining service life of the component, and identification of different approaches to minimize material degradation are useful investigations from the perspective of the long-term operations of industrial plants. The engineering material is subjected to different types of aging effects during its service life. Some examples are mechanical aging (fatigue and creep), thermal aging, environmental aging, hydrogen embrittlement, and aging due to nuclear irradiation. Engineering materials are required to have specified strength properties, such as yield stress, ductility, creep strength, fracture toughness, and hardness, to meet design requirements. These properties are generally met at the design stage by the selection of appropriate materials for the component. However, such properties of the selected materials are measured by testing newly manufactured engineering materials. Subsequently, when the materials are subjected to long-term operating conditions, they undergo microstructural changes. Such microstructural changes lead to the degradation of the mechanical properties. This reduces the design safety margins. Advanced experimental techniques are available to identify the degradation mechanisms of materials. Some advanced techniques are the X-ray diffraction technique, optical microscopy, scanning electron microscopy, atomic force microscopy, and transmission electron microscopy.

DOI: 10.1201/9781003310372-8

In addition to these experimental techniques, one can also use hybrid techniques to assess changes in the properties of degraded materials. Such hybrid techniques involve experimental procedures in association with analytical methods to quantify changes in the material properties. It is always desirable that such measurement techniques should be non-destructive as aged materials are obtained from real-life plant components. Such scrapping of materials from a plant component presently in service should not affect its integrity. One such technique is the small punch test (SPT) in association with empirical/semi-empirical correlations to analyze SPT data. The development of advanced correlations to analyze SPT data to calculate the material properties described in the previous chapters of the present book are useful for this purpose. These chapters, through case studies, present the procedures to develop such correlations. These procedures can also be employed by readers to develop similar correlations of other materials of their choice. The present chapter describes a real-life case study to evaluate the mechanical properties of a degraded material using experimental SPT data in association with correlations developed in earlier chapters. This will help readers to become acquainted with a step-by-step procedure for generating and utilizing SPT data to determine the properties of degraded materials useful for calculating updated safety margins and also to carry out life extension studies of plant components. In the present case study, oxygen-free electronic (OFE) copper is used as the candidate material. However, the procedure described is sufficiently generic in nature and can be applied to any other engineering material of the reader's requirements.

8.2 SELECTION OF THE CANDIDATE MATERIAL AND PROCEDURE ADOPTED TO OBTAIN DIFFERENT DEGREES OF MATERIAL DAMAGE

As described above, the purpose of the present chapter is to demonstrate a versatile methodology to assess changes in the mechanical properties of any material subjected to aging effects during its service life. For this purpose, OFE copper material is selected as the candidate material. This engineering material has wide applications, especially in nuclear industries. The scheme adopted to develop material degradation in OFE SPT specimens is nuclear irradiation. However, readers may note that the subsequent procedure to test SPT specimens and the analysis of the experimental data are general in nature and can be used for any degraded material, irrespective of the material damage mechanism.

Copper and copper alloys are versatile engineering materials. Some applications are for manufacturing bearings, gears and valve guides, radiators, hydraulic tubing and fasteners of pressure vessels, distillation equipment, piping systems and gaskets, and automotive parts. Copper alloys have unique combinations of physical properties, such as strength, corrosion resistance, conductivity, ductility, and machinability. These properties can be further enhanced by changing the chemical composition and also by adopting customized manufacturing techniques. The primary reason to select copper-based alloys is their high thermal conductivity. The high conductivity improves heat transfer to the coolant, allowing high heat fluxes on plant components without increasing the temperature of such components. This reduces thermal stress, which in turn improves the creep and thermal fatigue life of the components. The disadvantages of copper alloys are their relatively modest strength properties and low operating temperatures. One of the purest forms of an engineering material in this family is OFE copper, which has important applications in nuclear industries.

Nuclear irradiation during the service life of a nuclear reactor component alters the mechanical and fracture properties of the component material. It is mandatory to know such changes in the properties to quantify current safety margins. These data are also required to satisfy regulatory authorities to obtain a license for long-term plant operation. Nuclear irradiation leads to the formation of transmutation products. Irradiation also leads to the formation of defects, dislocation loops, stacking fault tetrahedra, and voids in the materials. This in turn alters the electrical and thermal conductivities, as well as the mechanical properties. Porosity swelling and transmutation contribute to a change in conductivity in some irradiated copper alloys. Irradiation significantly changes the work hardening behavior. It has been demonstrated that ductility is progressively reduced as the irradiation dose increases. Therefore, the irradiation process is selected in the present case study to develop damage in OFE copper material that leads to property degradation.

8.3 ELECTRON LINEAR ACCELERATOR (LINAC)

During the present case study, a 10 MeV LINAC shown in Fig. 8.1 and Fig. 8.2, is used to irradiate SPT specimens of OFE copper. The purpose of the irradiation is essentially to develop material damage in the SPT specimen of varying magnitudes. In the LINAC machine, an electron gun is the source of a 50 keV electron beam, which is further accelerated to 10MeV by oscillating electric fields driven by a microwave (or RF) power source. The irradiation of the material by the electron beam does not generate radioactive specimens in the case in which the electron energy is kept below 10 MeV. Due to this reason, the present accelerator is used to study the effect of material degradation and to assess changes in the mechanical properties. The SPT specimens are subjected to varying doses of irradiation to generate different magnitudes of material damage.

8.4 IRRADIATION OF SPT SPECIMENS OF OFE COPPER

As mentioned above, the SPT specimens are irradiated by electron beams with 10 MeV energy. Irradiation is carried out at a temperature of 300–350°C, one atmospheric pressure, and electron flux of 3×10^{14} cm^2/s. The specimens are mounted on a specially designed holder. Fig. 8.3 shows the initial raw material in the form of a plate used to fabricate the specimens, the fabricated SPT specimens, and a customized holder to mount the specimens during irradiation. A three-kilowatt

FIG. 8.1 Photograph of a 10 MeV accelerator used for the degradation study of OFE copper material

electron beam is used for irradiation. The temperature of the holder is measured during irradiation using k-type thermocouples with ±2.23°C accuracy. Temperature feedback from the recorder is used to control the temperature of the specimen. The nominal temperature is kept around 310°C. The power of the electron beam is regulated based on the measured temperature. Specimens are cooled by a forced circulation of air flow.

The irradiation time of the SPT specimens varies between 11 hours and 91.5 hours depending on the target dose. The dose rate of the accelerator is 1 MGy/hr/cm^2 at 3 kW beam power. For the beam parameters, the accelerated electrons can penetrate 38 mm thickness of a unit density material like water. For a copper material, it can penetrate up to 4.22 mm thickness. Hence, uniform damage across the thickness is expected in the present SPT specimens, which have thickness of the order of 250 microns. Multiple OFE copper specimens are subjected to electron irradiation up to 11, 46, 62.5, 80.5, and 91.5 MGy. All the irradiated specimens are tested to obtain load vs. punch displacement data up to breaking point in a specialized machine available in the Miniaturized Specimen Test lab of the Bhabha Atomic Research Centre.

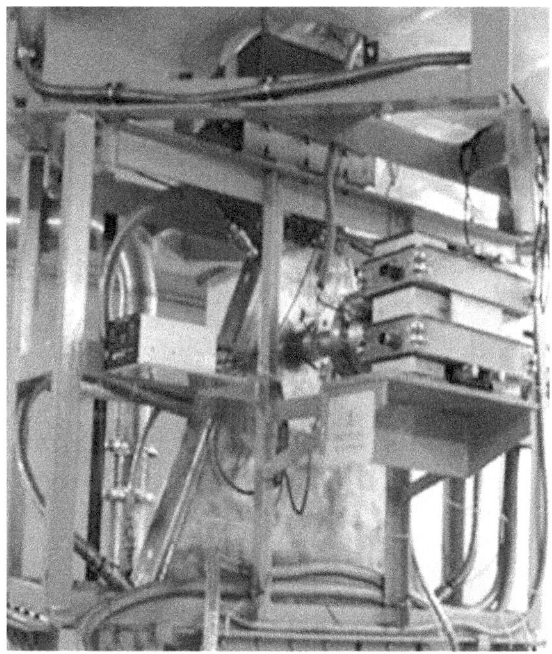

FIG. 8.2 Photograph of the irradiation area of a 10 MeV accelerator used in the present case study

8.5 PROCEDURE TO TEST DEGRADED SPT SPECIMENS AND ANALYSIS OF THE RESULTS

8.5.1 Generation of Load vs. Punch Displacement Data

The plots of the measured load vs. punch displacement data for all the SPT specimens are shown in Fig. 8.4. The specimens are subjected to five doses of irradiation: 11, 46, 62.5, 80.5, and 91.5 MGy. In some cases, multiple specimens are also available, depending on the availability of the accelerator machine. As mentioned earlier, the cumulative irradiation dose in the MGy unit is the quantitative representation of material damage in each specimen. Material hardening is observed beyond 62.5 MGy, as shown in Fig. 8.4. This is indicated by the shift in fracture points toward lower values of punch displacements. Fig. 8.5 and Fig. 8.6 show scanning electron microscopy (SEM) images of the fracture surfaces of two SPT specimens. The load vs. punch displacement data of the SPT specimens shown in Fig. 8.4 are further processed to determine changes in the different mechanical properties of OFE copper material as a function of material damage expressed in the MGy unit. The details are given in the subsequent subsections.

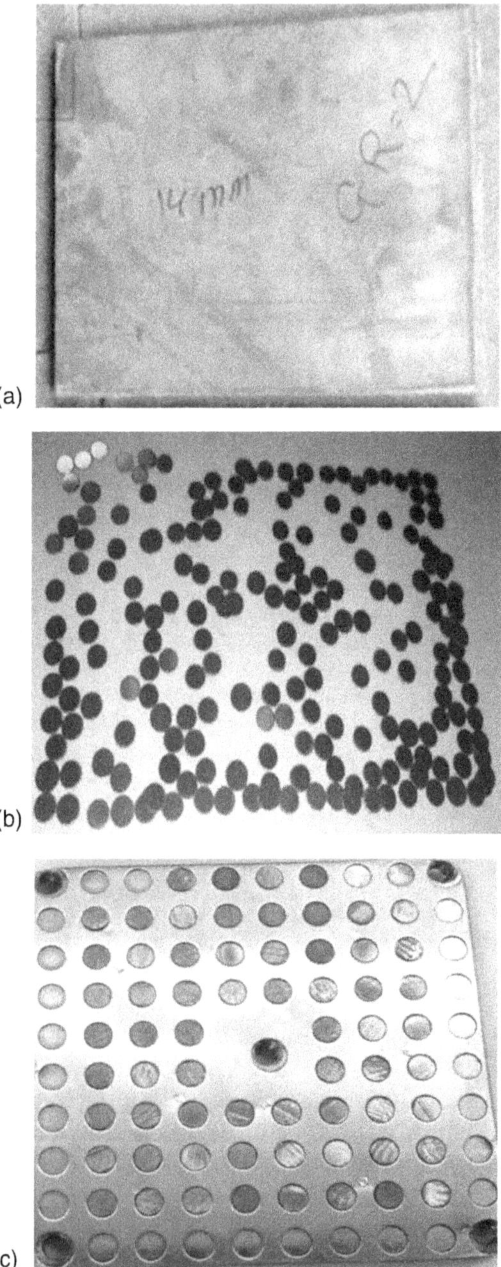

FIG. 8.3 Photographs of (a) raw material used to manufacture SPT specimens, (b) multiple SPT specimens prepared for irradiation, and (c) mounting arrangement of SPT specimens in the accelerator for irradiation by an electron beam

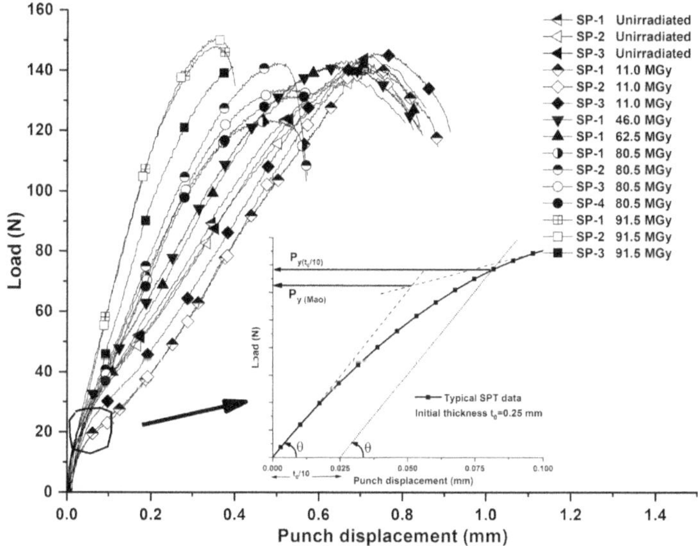

FIG. 8.4 Load vs. punch displacement plots of SPT specimens of OFE copper material subjected to electron irradiation. The material damage is expressed in the MGy unit

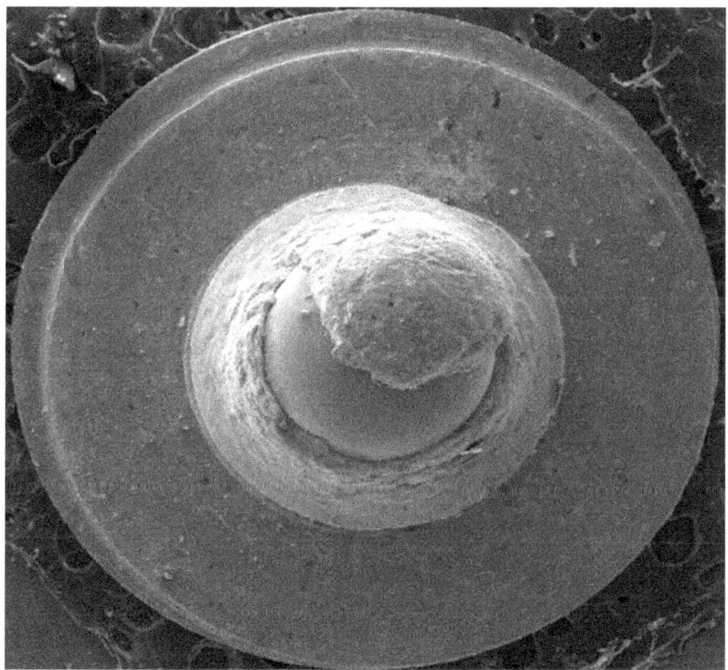

FIG. 8.5 SEM image of the fracture of an SPT specimen along with the spherical punch

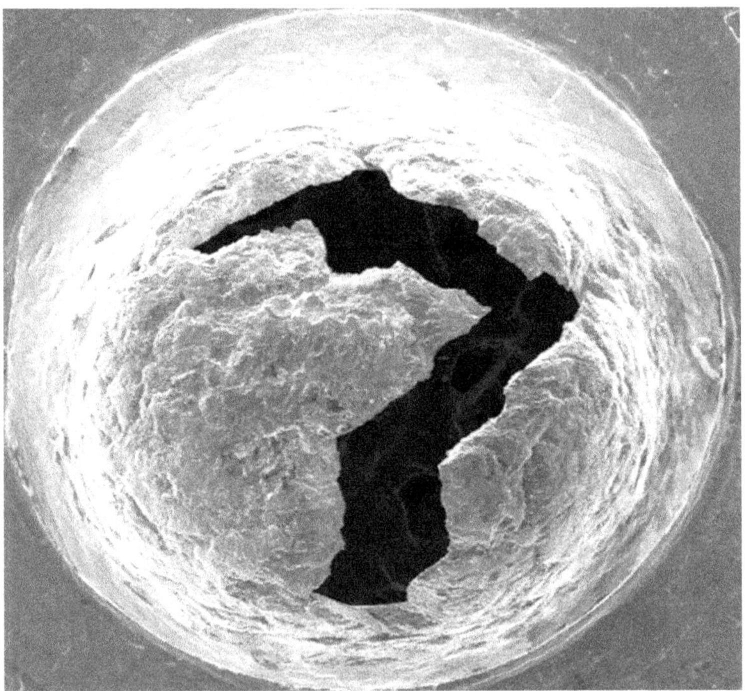

FIG. 8.6 SEM image of the fracture profile of an SPT specimen after testing

8.5.2 Determination of Yield and Ultimate Stresses

The analytical development of new correlations to evaluate the yield and ultimate stresses of copper alloys using SPT data are presented in Chapter 3. Following the same procedure, two independent correlations can be derived to calculate the yield stress of copper alloys. The input to the first correlation is $P_{y(Mao)}$. This parameter is determined by calculating the load at the intersection point between the two tangents drawn in the elastic and plastic regions, as shown in Fig. 8.4 [1]. The input to the second correlation $P_{y(}t_{0/10)}$ is the load at the intersection point between the experimental SPT curve and a straight line drawn parallel to the initial slope of the SPT curve with an offset displacement of $t_0/10$. This methodology is also shown in Fig. 8.4. These correlations are

$$\sigma_{YS} = 0.3586\ P_{y(Mao)}/t_0^{\ 2} \tag{8.1a}$$

$$\sigma_{YS} = 0.2650\ P_{y(t0/10)}/t_0^{\ 2}. \tag{8.1b}$$

The correlation to determine σ_{UTS} of copper alloys using SPT data as derived in Chapter 3:

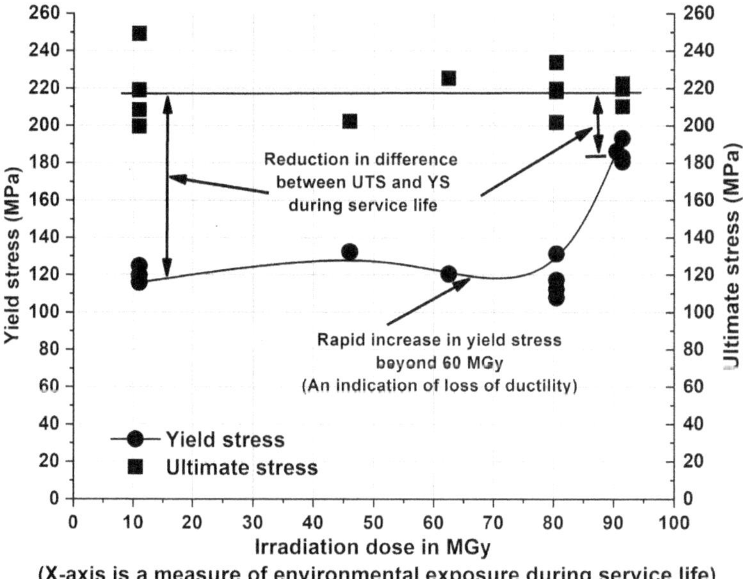

FIG. 8.7(a) Variations of YS and UTS as a function of material damage expressed in terms of the MGy unit of the irradiation dose

$$\sigma_{UTS} = 0.11315\, P_{max}/t_0^{\,2} - 52.435. \tag{8.2}$$

In all these correlations, t_0 is the initial thickness of the SPT specimen (mm), σ_{YS} is the yield stress (MPa) and σ_{UTS} is the ultimate tensile stress (MPa) of the material.

The load vs. punch displacement plots of the SPT data are applied to calculate $P_{y(Mao)}$, and $P_{y(t0/10)}$. Some difficulties are encountered in obtaining a consistent value of $P_{y(Mao)}$ for all the plots in Fig. 8.4. To overcome this difficulty, the values of $P_{y(t0/10)}$ are used to calculate the yield stress using eq. (8.1b). Eq. (8.2) is employed to calculate σ_{UTS} by determining the P_{max} of all the curves in Fig. 8.4. Plots of the yield stress and the ultimate stress as a function of the material damage parameter represented in MGy are shown in Fig. 8.7(a). Readers may note that wherever multiple specimens are available for a particular value of MGy, the average values of the yield and ultimate stresses are used to show the graphical variations in Fig. 8.7 (a).

8.5.3 Determination of the 'Biaxial Strain' and 'Biaxial Strain at Fracture'

As explained in Chapter 2, an expression of the equivalent biaxial strain (bulge ductility) was proposed in [2] using membrane theory. This expression is

$$\varepsilon_q = \ln(t_0/t), \tag{8.3}$$

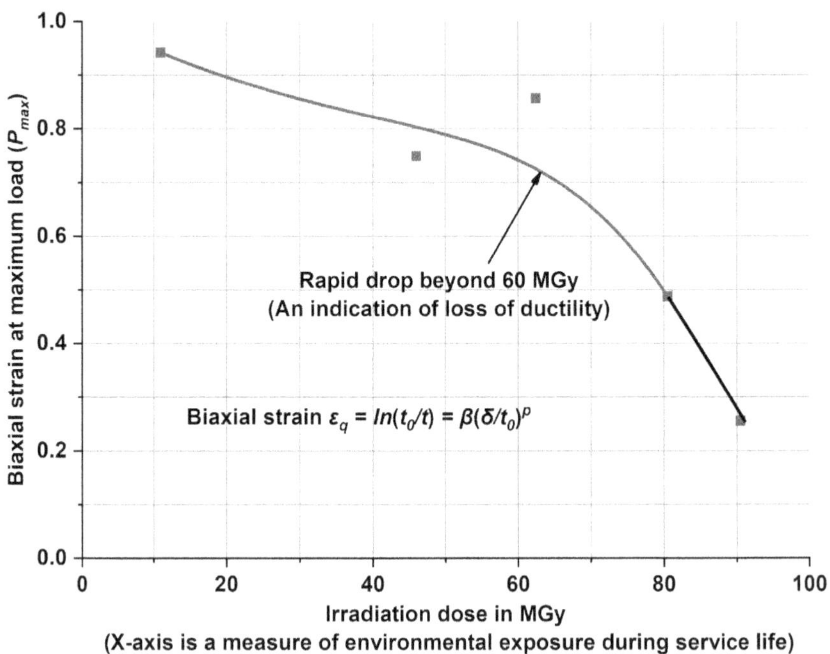

FIG. 8.7(b) Variation of the biaxial strain at the maximum load as a function of the degree of material degradation expressed in terms of the unit of MGy of the irradiation dose

where t is the instantaneous minimum thickness during the experiment and t_0 is the initial thickness of the SPT specimen. To make use of eq. (8.3), one needs to know the instantaneous minimum thickness during the experiment. The measurement of such a parameter is difficult and susceptible to considerable errors. To circumvent this problem, an equation correlating the instantaneous central displacement (δ) of the specimen with the corresponding instantaneous minimum thickness (t) is suggested in [2]. This correlation along with eq. (8.3) leads to the following correlation of the biaxial strain:

$$\varepsilon_q = \ln(t_0/t) = \beta(\delta/t_0)^p. \tag{8.4}$$

Eq. (8.4) can be used to calculate the biaxial strain at any instant during the experiment of SPT specimens by using load vs. punch displacement data. By substituting the value of the punch displacement at fracture (δ_f), the same expression can also be used to calculate the biaxial fracture strain (ε_{qf}). It is shown in Chapter 4 that the parameters β and p in eq. (8.4) depend on the initial thickness of the SPT specimen. Using the expressions suggested in Chapter 4 and the results shown in Fig. 8.4, the biaxial strains at maximum load are calculated for all the SPT specimens. A plot of these values is shown in Fig. 8.7(b). In addition, the biaxial fracture strains are also calculated using punch displacements at the fracture points of all the specimens. A plot of this parameter is also shown in Fig. 8.8.

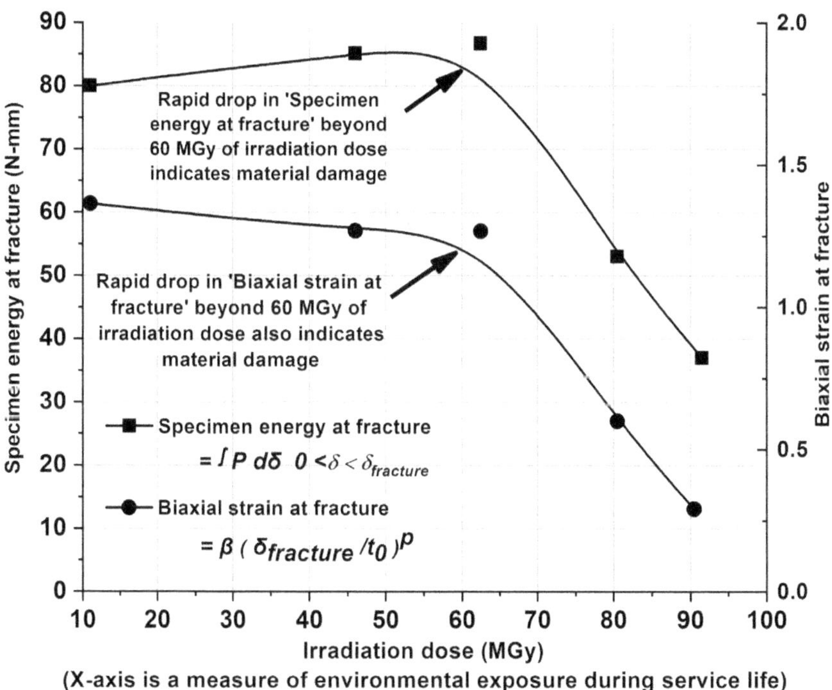

FIG. 8.8 Variation of the specimen energy and biaxial strain at fracture as a function of material degradation expressed in the MGy unit of the irradiation dose

8.5.4 Determination of Specimen Energy at Fracture

The specimen energy is the strain energy stored in the SPT specimen during the experiment. This is measured by calculating the area under the load vs. punch displacement curve using

$$SE = \int P \, d\delta, \tag{8.5}$$

where P is the instantaneous punch load and δ is the corresponding punch displacement.

Eq. (8.5) is applied to calculate the specimen energy at fracture of all the SPT specimens using the experimental results shown in Fig. 8.4. These values are also plotted in Fig. 8.8. Fig. 8.9 shows a plot between the biaxial fracture strain and specimen energy at fracture for all the SPT specimens. The equation of the best fit linear curve derived from this plot is

$$\varepsilon_{qf} = 0.02155 S_e - 0.50687, \tag{8.6}$$

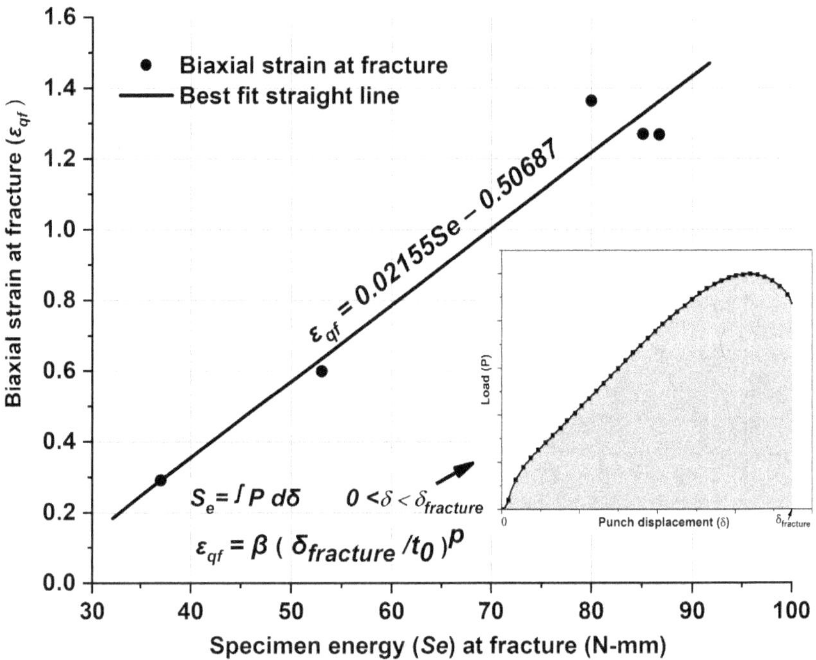

FIG. 8.9 Variation of the biaxial fracture strain with the specimen energy at fracture

where ε_{qf} is the biaxial fracture strain and S_e is the specimen energy at fracture in N-mm. Readers may note that the empirical correlation shown in eq. (8.6) provides a more convenient way to calculate the biaxial fracture strain. However, its material dependency and unit dependency should be kept in mind while using this equation.

8.5.5 Generation of the Entire Stress-Strain Curve of Degraded OFE Copper by Evaluating Holloman Power Law Constants

The methodology to calculate the yield stress and ultimate stress of a degraded material by using SPT data is shown in Section 8.5.2. It is always desirable to generate the entire stress-strain data of a degraded material. The present section demonstrates such a methodology using the results shown in Fig. 8.4. This is done by employing the Holloman power law to represent the entire material stress-strain data. The expression of the Holloman power law is

$$\sigma = \sigma_{YS} + K\varepsilon_p^n, \tag{8.7}$$

where σ_{YS} is the true stress, ε_p is the true plastic strain, and K and n are the empirical constants. To obtain baseline data, it is desirable to find the K and n

TABLE 8.1
Mechanical properties of OFE copper before putting it into service

Material	E (MPa)	σ_{YS} (MPa)	σ_{UT} (MPa)	K (MPa)	n
OFE Copper	115000	141	206	317.8	0.159

of the fresh material before any degradation. Such values can be obtained either from the literature or by conducting experiments on tensile specimens. Table 8.1 shows the properties of OFE copper before any material is put into service.

The material properties shown in Table 8.1 are then further used to identify the Holloman power law constants of the degraded OFE copper. For this purpose, a parametric study is carried out using a finite element (FE) model of the SPT specimen, along with the appropriate boundary conditions. For the numerical analysis, the material stress-strain curve is represented by eq. (8.7) and the yield stress is taken from Fig. 8.7(a) as a function of material degradation expressed in terms of the MGy unit. The parametric study is conducted by varying the Holloman power law constants with an objective to numerically match the calculated load vs. punch displacement curves as closely as possible with the experimental results of Fig. 8.4 up to the maximum load. Figs. 8.10 (a–d) show the numerically calculated SPT load-displacement curves in comparison with the experimental results for the most optimum set of Holloman power law constants. These figures show the results of fresh OFE copper before any degradation and also for OFE copper degraded by irradiation doses of 62.5 MGy, 80.5 MGy, and 91.5 MGy, respectively. The optimum sets of K and n are shown in Table 8.2. The sections of the experimental plots used to ascertain K and n in parametric studies, are shown in respective figures using arrow marks. No attempt has been made to generate Holloman power law constants for intermediate irradiation doses of 0–62.5 MGy as there is no noticeable change in ductility, as indicated by Fig. 8.7(a). Hence, no change in the Holloman power law constants is expected over this range of doses. The following two observations may be drawn from Figs. 8.10(a–d):

a. Over the range 0–62.5 MGy, it is possible to obtain a set of Holloman power law constants that compute load vs. punch displacement data close to the experimental data up to the maximum load. It may be noted that the present FE elastic-plastic analyses do not have the ability to model crack initiation and propagation in the SPT specimen in the absence of any numerical material damage model, that is, the Gurson model. The elastic-plastic analysis results have good matching with the experimental results over this range of irradiation dose, which indicates no crack initiation in the SPT specimen up to the maximum load between 0.0 MGy and 62.5 MGy. This is expected as there is no loss of material ductility over this range of irradiation dose, as mentioned earlier.

TABLE 8.2
Flow properties of degraded OFE copper material determined by FE parametric studies of experimental results

	YS (MPa)	*n*	*K* (MPa)
Unirradiated	141	0.159	317.8
62.5	120	0.14	350
80.5	136	0.1	400
91.5	180	0.05	450

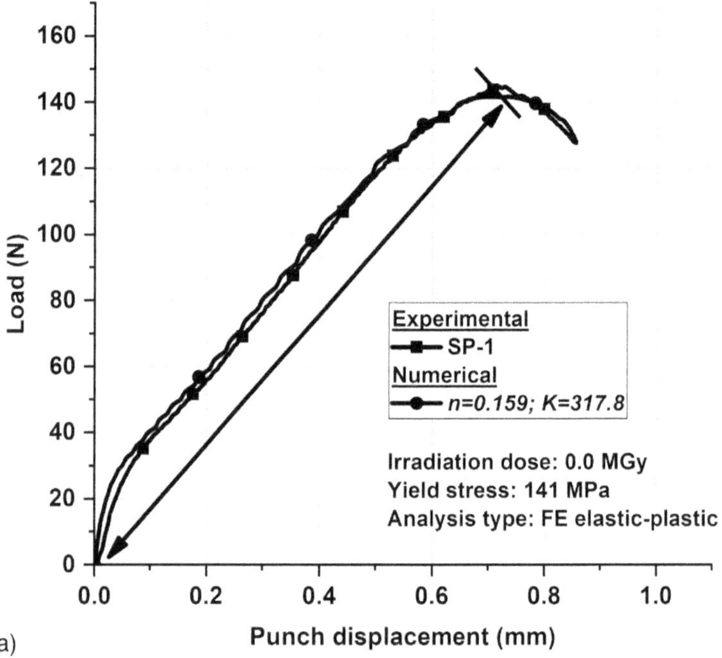

FIG. 8.10 Parametric studies to ascertain the Holloman power law constants for different levels of material degradation by irradiation: (a) unirradiated, (b) 62.5MGy, (c) 80.5MGy, and (d) 91.5MGy

b. For the irradiation doses of 80.5 MGy and 91.5 MGy, the numerically calculated load vs. punch displacement curves match well only with the initial sections of the experimental data for an optimum set of *K* and *n*. However, beyond this section, the experimental load-displacement curves show the failure of the SPT specimens before attaining the maximum load calculated by elastic-plastic

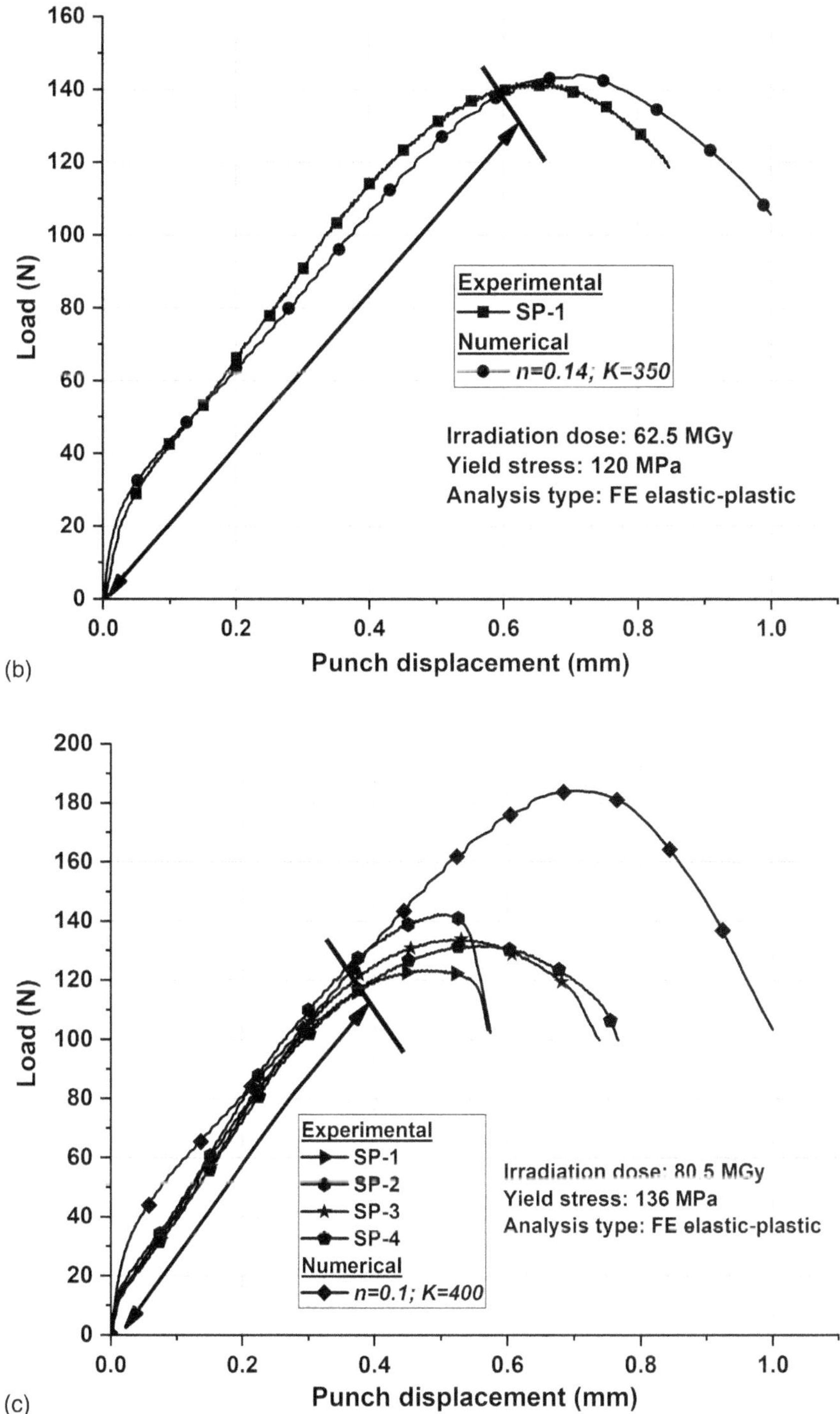

FIG. 8.10 (Continued)

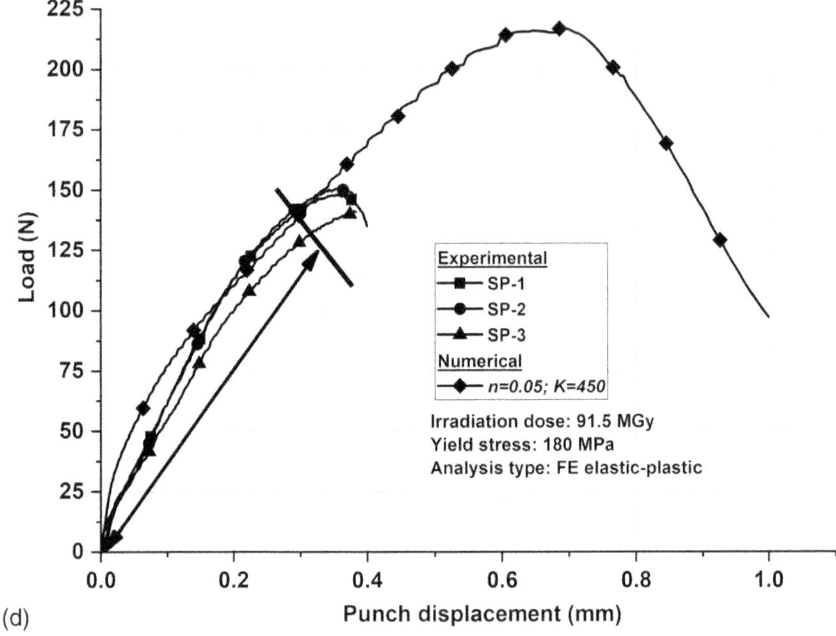

FIG. 8.10 (Continued)

analysis. This indicates that the material is in a semi-brittle state and crack propagation occurs before the maximum load is attained during the experiment. This signifies a loss of ductility in OFE copper over this range of irradiation dose. This is also visible in Figs. 8.7(a) and (b). The optimum sets of material properties for different levels of irradiation dose determined through the present parametric studies are shown in Table 8.2.

8.5.6 Elastic-plastic Analysis along with Cohesive Zone Elements to Model the Semi-Brittle Characteristic of Degraded OFE Copper

As mentioned above, elastic-plastic analyses of SPT specimens are unable to match the experimental load-displacement data of SPT specimens that are subjected to cumulative degradation equivalent to irradiation doses of more than 62.5 MGy. This is primarily due to a loss of ductility of OFE copper beyond this dose. Over this range, experimental peak loads are found to be significantly less than the peak loads calculated using elastic-plastic FE analysis. These observations are also noted in Figs. 8.10(c) and (d). Such an observation over this range of irradiation is due to crack growth before the specimen fracture load is attained. Elastic-plastic analysis is not able to model such crack growth phenomena.

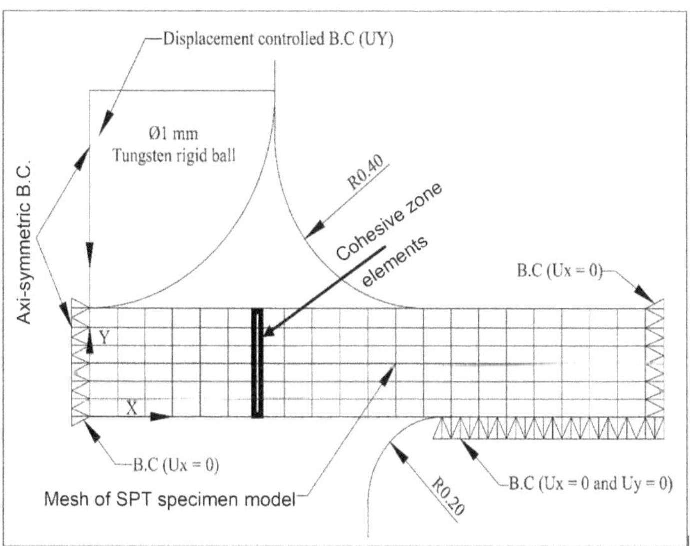

FIG. 8.11 SPT specimen mesh with a layer of cohesive zone elements to model crack initiation and propagation as observed during the experiment

To overcome this difficulty, it is necessary to incorporate a damage mechanics module in the FE code. Such a module helps to model crack initiation and propagation in the SPT specimen when carrying out elastic-plastic analysis. Readers are acquainted with the *Gurson damage mechanics model* widely used to model crack initiation and propagation, as explained in Chapter 7. In the present chapter, another damage mechanics model called *cohesive zone modeling* is employed. Readers are advised to go through the voluminous literature available on the formulations and applications of cohesive zone modeling. The FE model of the SPT specimen is updated by introducing a thin layer of cohesive zone elements along the thickness. Fig. 8.11 shows the location of the cohesive zone elements schematically. The radial location of this thin layer of cohesive zone elements is appropriately chosen so as to match the location of the fracture in the SPT specimen as observed during the experiment.

The input to cohesive zone modeling is a traction–separation law (TSL) [3, 4]. In the present study, the shape of the TSL is chosen to be either bilinear or trapezoidal depending on the degree of material degradation. Fig. 8.12(a) shows the shape of the trapezoidal TSL and the associated parameters. Fig. 8.12(b) shows the values of the material parameters used in the present analysis of OFE copper. It may be seen that three parameters, that is, maximum traction (T_0), damage ratio (r), and cohesive energy (Γ_0) are assumed to be varying as a function of material degradation represented by the irradiation dose. These parameters are subsequently

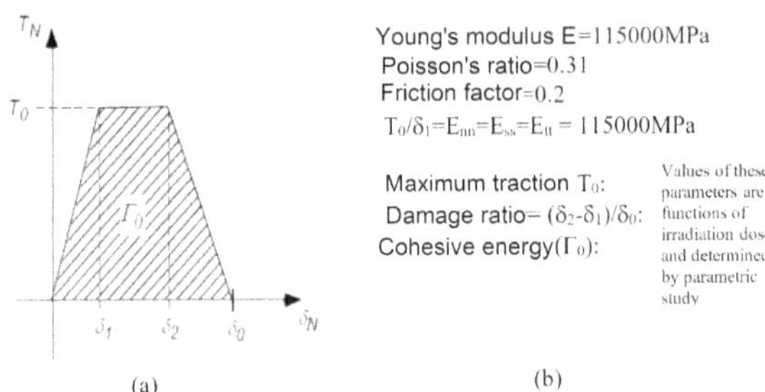

FIG. 8.12 (a) Description of the trapezoidal TSL and associated parameters, and (b) material parameters used in the present analysis

TABLE 8.3
Holloman power law constants and TSL parameters ascertained by carrying out parametric studies of irradiated OFE copper SPT specimens

	Holloman power law constants			Cohesive zone TSL parameters		
Material degradation expressed in terms of irradiation dose	***n***	***K* (MPa)**	***YS* (MPa)**	**T_{max} (MPa)**	**Cohesive energy (N/mm)**	**Damage ratio *r***
Unirradiated	0.159	317.8	141	230	18	0
62.5 MGy	0.15	355	120	248	15.2	0
80.5 MGy	0.1	400	136	268	15	0.2
91.5 MGy	0.05	450	180	225	13	0.47

determined by carrying out an FE parametric study. The parametric studies are carried out to obtain the best set of these parameters that generates numerical load vs. punch displacement data close to the experimental results shown in Fig. 8.4. Table 8.3 shows the optimum values of these parameters as a function of irradiation doses. It may be seen that the '*r*' parameter is found to be negligible up to a dose of 62.5 MGy. This signifies the suitability of a bilinear TSL for these cases. However, for higher values of doses, that is, 80.5 MGy and 91.5 MGy, parameter '*r*' is found to be non-zero, signifying the suitability of trapezoidal TSL for this range of doses. Such a change in the shape of the TSL is expected due to the loss of ductility of OFE copper for higher values of irradiation doses, and the material behaves in a semi-brittle manner. Fig. 8.13(a) shows a comparison of experimental load vs. punch displacement data with the results of FE analysis using cohesive zone elements for the irradiation doses of 0.0, 62.5 MGy, 80.5 MGy, and 91.5 MGy. Fig. 8.13(b) shows the optimum shapes of the TSL used in these cases. Fig. 8.14 shows the typical progress of damage along the cohesive layer of the specimen

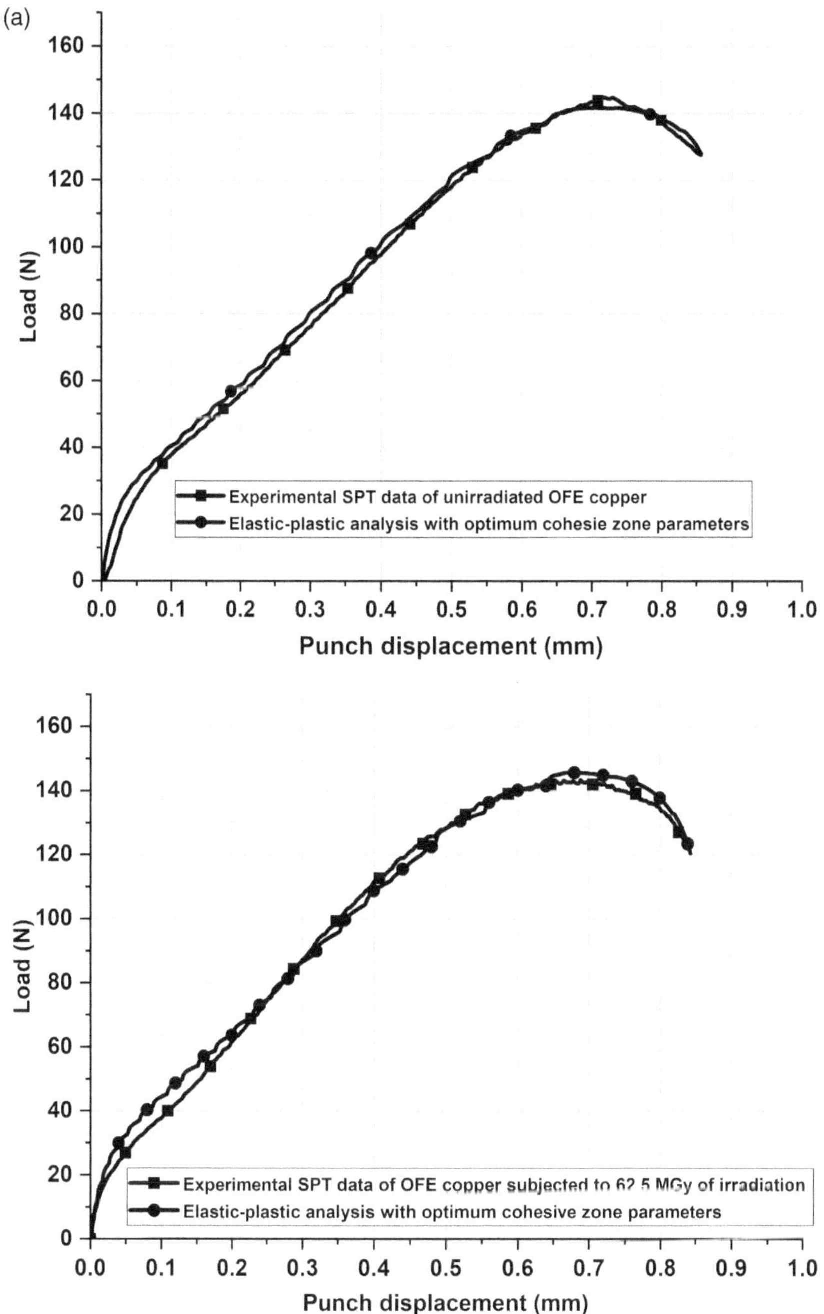

FIG. 8.13 (a) Comparison of experimental load-displacement data with elastic-plastic FE analysis applying optimum cohesive zone parameters, and (b) shape of TSL cohesive zone elements with optimum parameters applied for unirradiated, 62.5 MGy, 80.5 MGy, and 91.5 MGy of the irradiation doses

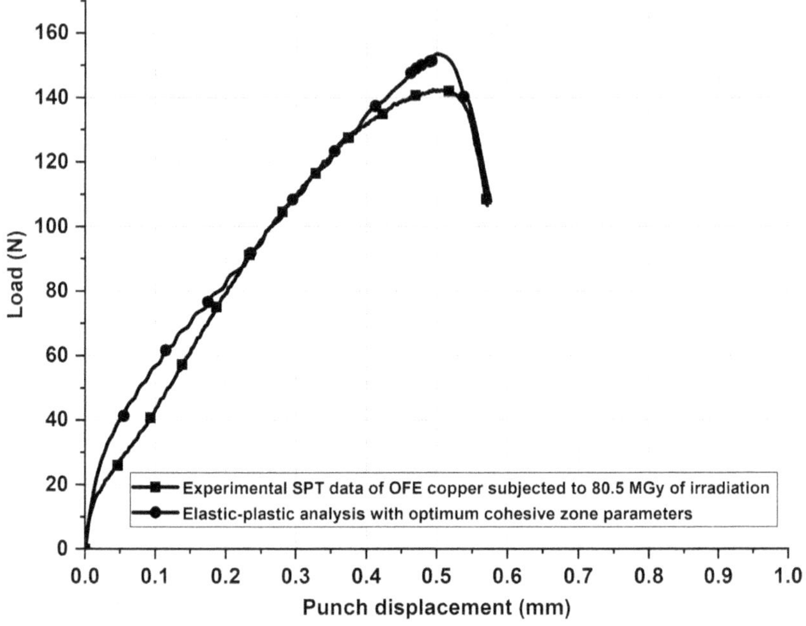

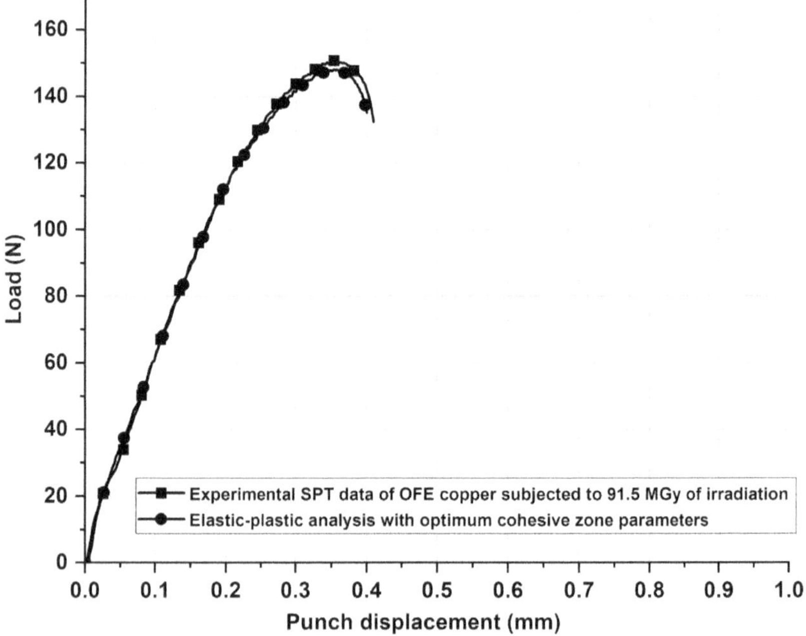

FIG. 8.13 (Continued)

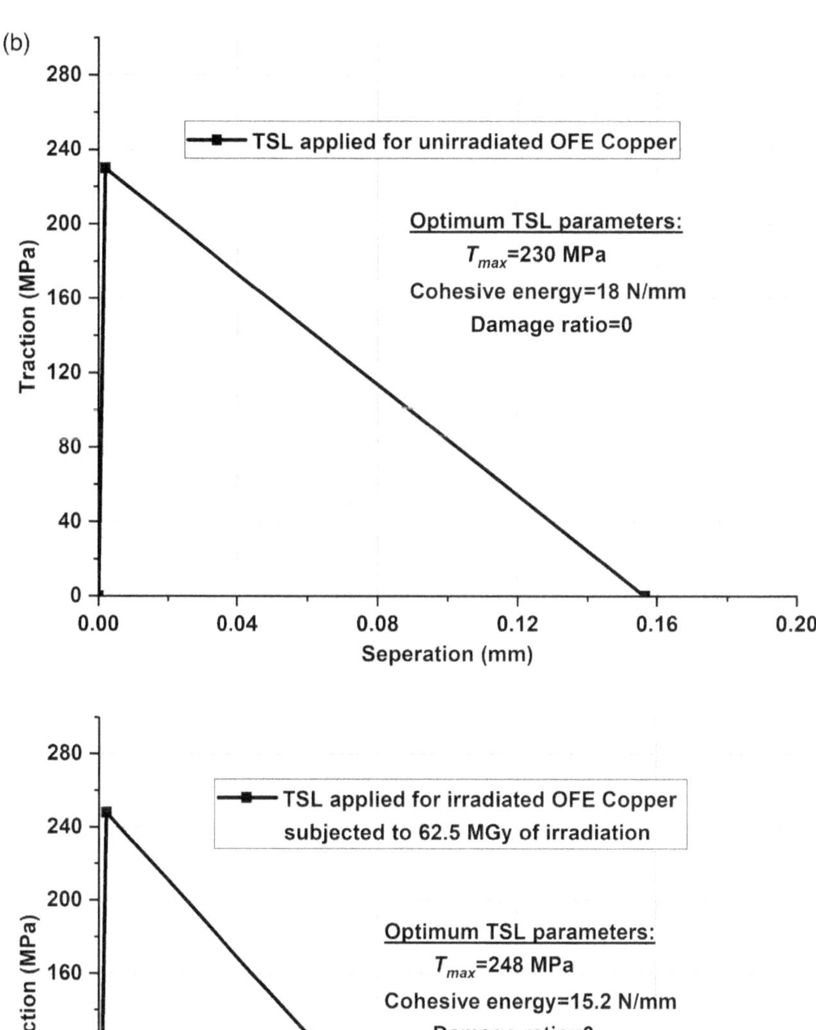

FIG. 8.13 (Continued)

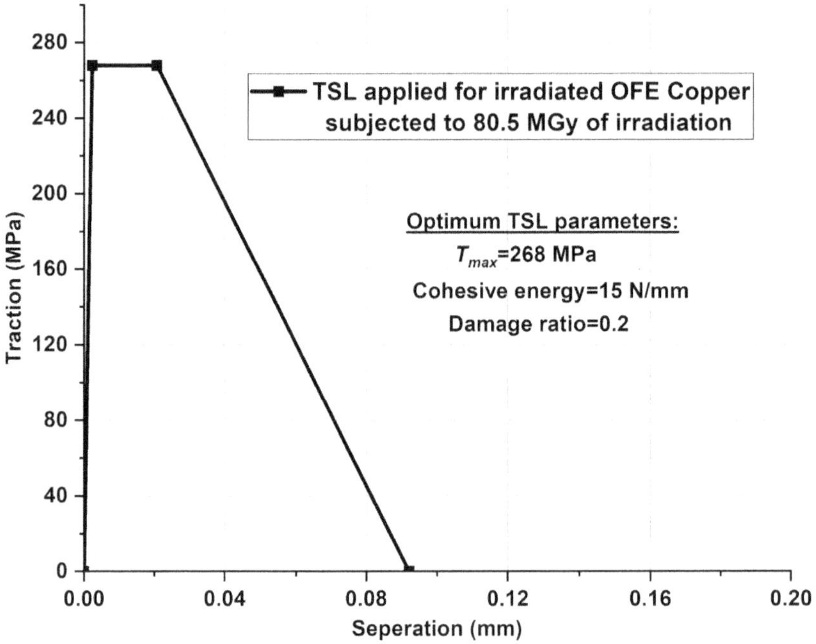

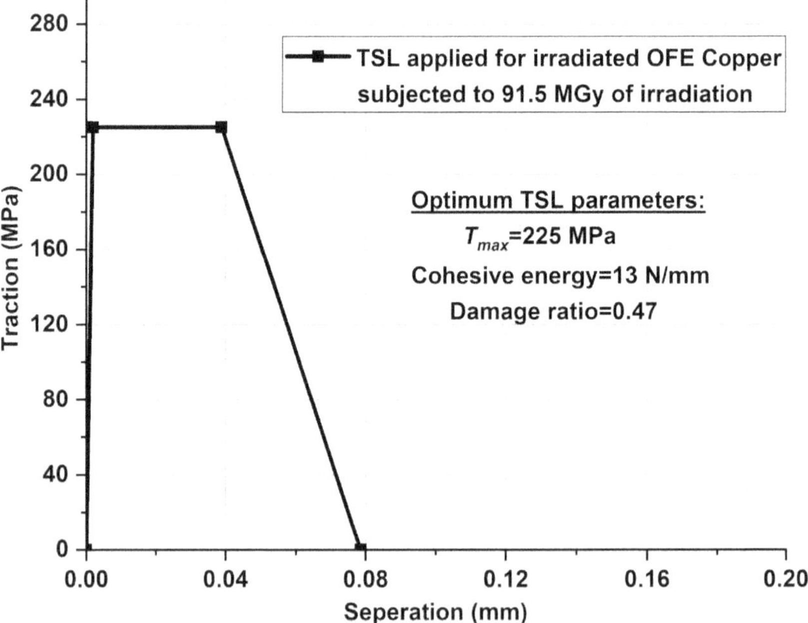

FIG. 8.13 (Continued)

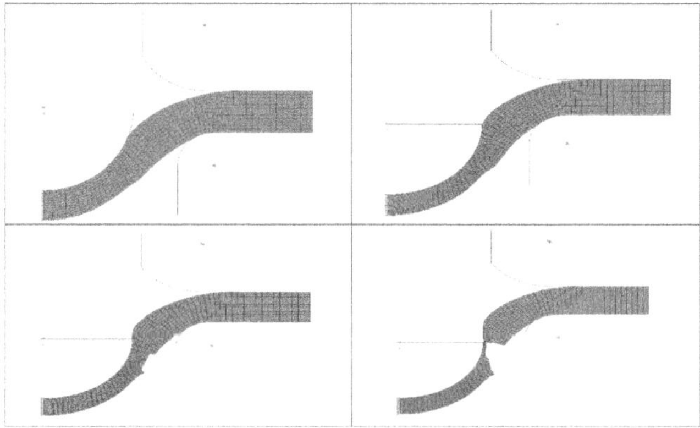

FIG. 8.14 FE analysis of the SPT specimen of OFE copper material with cohesive zone elements. These figures show the typical progress of damage along the specimen thickness leading to the fracture of the specimen

thickness leading to the fracture of the specimen. Readers may note that Table 8.3 provides a complete set of material parameters to carry out the safety analysis of any industrial component made up of OFE copper subjected to different degrees of material degradation due to irradiation.

8.5.7 A Methodology to Calculate the Fracture Toughness of Degraded OFE Copper

The material parameters shown in Table 8.3 can be used to calculate the fracture toughness of OFE copper subjected to degradation. For this purpose, an FE model of a three-point bending (TPB) specimen of ASTM standard is prepared. The schematic diagram of the specimen is shown in Fig. 8.15(a). The dimensions of the specimen are 112.86 mm × 6.76 mm × 25.08 mm (L × B × W) and the initial crack length is a_0 = 7.6494 mm. Hence, a_0/W is 0.3. The FE model is shown in Fig. 8.15(b). The two-dimensional plane strain four-noded iso-parametric elements with appropriate boundary conditions are used in the model. The cohesive zone elements are placed along the remaining ligament of the specimen starting from the crack tip. Such a modeling strategy enables the consideration of crack initiation and propagation with elastic-plastic analysis due to material damage.

FE analyses are carried out for irradiation doses of 0.0, 62.5, 80.5, and 91.5 MGy separately. The elastic-plastic material properties and TSL parameters shown

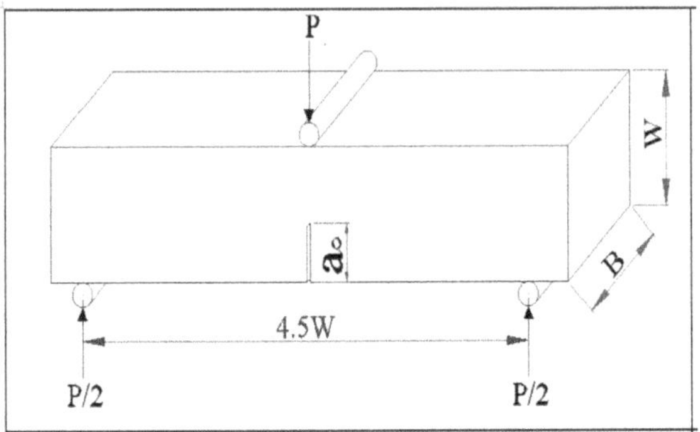

FIG. 8.15(a) Schematic diagram of a TPB specimen analyzed to calculate the J-R curve

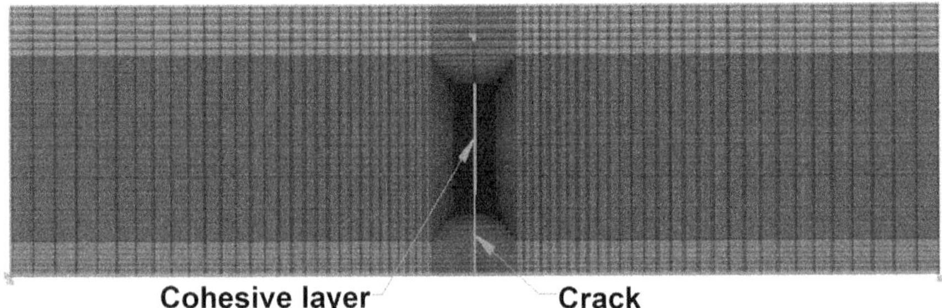

FIG. 8.15(b) FE model of the TPB specimen with a single-edge crack. Cohesive zone elements are placed along the remaining ligament to model crack initiation and propagation

in Table 8.3 are applied. The outputs of the FE analysis are load vs. CMOD and crack growth vs. CMOD data. The load vs. CMOD data are then converted to J-integrals using the procedure described in ASTM E1820 [5]. The sample plots of load vs. CMOD and J-integral vs. crack growth for 80.5 MGy doses of irradiation are shown in Fig. 8.16. The J-integral vs. crack growth data of all the doses of irradiation are used to calculate J-initiation by extrapolating the J-R curve to the *y*-axis, that is, zero crack growth. Table 8.4 shows the values of J-initiation and Fig. 8.17 shows the variation of J-initiation as a function of the doses. The following linear equations derived from Fig. 8.17 may be used to interpolate J-initiation for intermediate values of irradiation doses Φ:

$$\text{J-initiation} = 30.92 - 0.0840\Phi \qquad 00.0 \le \Phi \le 62.5$$

$$\text{J-initiation} = 25.67 - 0.2917(\Phi\text{-}62.5) \qquad 62.5 < \Phi \le 80.5$$

$$\text{J-initiation} = 20.42 - 0.6072(\Phi\text{-}80.5) \qquad 80.5 < \Phi \le 91.5,$$

where J-initiation is in N/mm and irradiation dose Φ is in MGy.

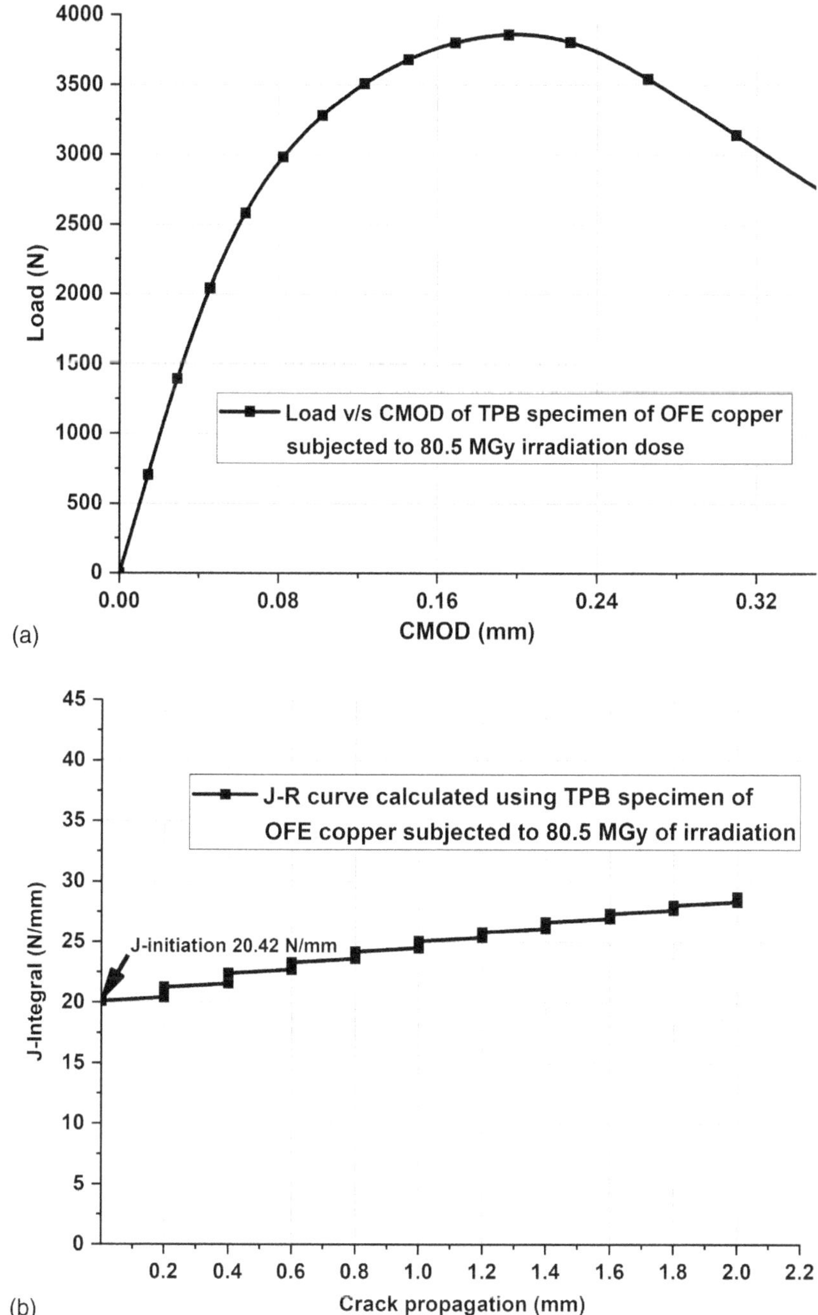

FIG. 8.16 Sample plot of (a) load vs. CMOD and (b) J-R curve of a TPB specimen of OFE copper material subjected to 80.5 MGy of irradiation dose

TABLE 8.4
Fracture toughness of OFE copper as a function of material degradation due to irradiation dose

Dose (MGy)	*Ji* (N/mm)
00	30.92
62.5	25.67
80.5	20.42
91.5	13.74

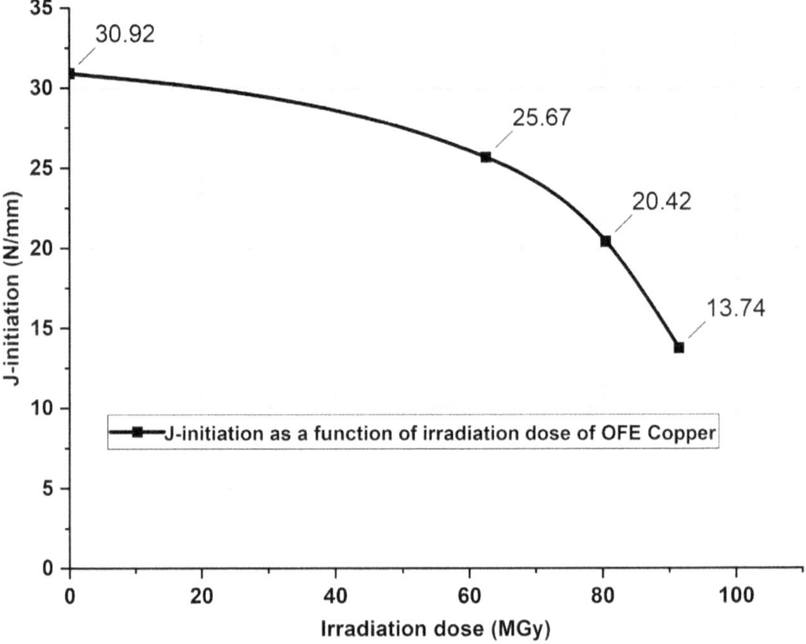

FIG. 8.17 Variation of J-initiation as a function of material degradation represented by the irradiation dose in the MGy unit

8.6 CHAPTER CLOSURE

1. The present chapter deals with a case study to determine the degraded properties of a material using SPT specimens. The eventual applications of the outcome of the previous chapters of the present book are demonstrated in this case study.

2. Due to the miniaturized size of the SPT specimens, a small amount of degraded material can be scrapped from industrial components without affecting their serviceability. Such materials may be used to assess the extent of degradation as a function of the service period. As component integrity is not affected by scrapping a small amount of aged material, such a technique may be termed non-destructive examination.
3. For the sake of the present case study, the candidate material selected is OFE copper and the procedure adopted to cause material damage is nuclear irradiation. However, readers may note that the methodology adopted to determine changes in the material properties is generic and can be used for any other degraded material and any other damage mechanism.
4. The candidate OFE copper material has wide applications in the nuclear industry, especially in fusion reactors and accelerators. The mechanical properties, such as yield stress, ductility, fracture toughness, fatigue, and creep, are affected when this material is subjected to nuclear irradiation during its service life.
5. To carry out the present case study, 15 SPT specimens of OFE copper are subjected to electron irradiation up to 11, 46, 62.5, 80.5, and 91.5 MGy to cause material damage in the specimens. A linear accelerator is used for this purpose. The irradiated specimens are then tested to obtain load vs. punch displacement data.
6. Changes in the yield stress, ultimate tensile stress, biaxial fracture strain, specimen energy at fracture, and fracture initiation toughness as a function of material degradation represented by the irradiation dose are then calculated using experimental SPT specimen data and employing empirical correlations described in earlier chapters of the present book.
7. A noble hybrid method of applying cohesive zone modeling along with elastic-plastic analysis is presented to quantify the fracture initiation toughness (J_i) and fracture resistance data (J-R curve). An FE parametric study is carried out to identify the material cohesive parameters by comparing computed load vs. punch displacement data with the measured values.
8. It is seen that the loss of ductility of OFE copper is significant beyond 62.5 MGy of irradiation. Such a conclusion is consistently observed while calculating changes in the yield stress, ultimate tensile stress, biaxial fracture strain, specimen energy at fracture, and fracture initiation toughness as a function of material degradation. This shows the inherent strength of the SPT methodology to generate consistent and reliable quantitative information.
9. It is expected that the advanced methodology presented in this book to evaluate changes in the properties of aged materials during the service life using SPT data are useful to designers for safety evaluation and also

to calculate the remaining service life of industrial components for life extension studies.

REFERENCES

[1] X. Mao, H. Takahashi, Development of a further miniaturized specimen of 3 mm diameter for TEM disk (φ 3 mm) small punch tests. Journal of Nuclear Materials 150 (1987) 42–52.

[2] J. Chakrabarty, A theory of stretch forming over hemispherical punch heads. International Journal of Mechanical Sciences 12 (1970) 315–325.

[3] K. Schwalbe, I. Scheider, A. Cornec, Guidelines for Applying Cohesive Models to the Damage Behavior of Engineering Materials and Structures. Springer Briefs in Applied Sciences and Technology (2013) ISBN: 9783642294938.

[4] I. Scheider, Cohesive model for crack propagation analyses of structures with elastic-plastic material behavior Foundations and implementation. (2001) GKSS research center Geesthacht, Dept. WMS, 2001.

[5] ASTM Standard E1820-18a, Standard Test Method for Measurement of Fracture Toughness. (2019) Annual Book of ASTM Standards, American Society for Testing and Materials. DOI: 10.1520/E1820-18

Index

Note: Page numbers in *italic* refer to figures, page numbers in **bold** refer to tables.

For Product Safety Concerns and Information please contact our EU representative GPSR@taylorandfrancis.com
Taylor & Francis Verlag GmbH, Kaufingerstraße 24, 80331 München, Germany

www.ingramcontent.com/pod-product-compliance
Lightning Source LLC
LaVergne TN
LVHW020717110826
845149LV00012B/2306

* 9 7 8 1 0 3 2 3 1 5 7 6 8 *